Uses of Inorganic Chemistry in Medicine

Uses of Inorganic Chemistry in Medicine

Edited by

Nicholas P. Farrell
Virginia Commonwealth University, Richmond, Virginia, USA

ROYAL SOCIETY OF CHEMISTRY

Cover illustration by Philip Mattes

ISBN 0-85404-444-2

A catalogue record for this book is available from the British Library.

Published by The Royal Society of Chemistry,
Thomas Graham House, Science Park, Milton Road, Cambridge CB4 0WF, UK

For further information see our web site at www.rsc.org

Typeset in Great Britain by Vision Typesetting, Manchester
Printed and bound by MPG Books Ltd, Bodmin, Cornwall, UK

Preface

Any compilation of clinically used drugs must include important inorganic pharmaceuticals such as lithium carbonate and cisplatin. Studies in the field of inorganic-based pharmaceuticals continue to expand; while yet representing a relatively small percentage of total drugs, the inorganic-based drugs fulfil important roles. New developments in metal-based drugs must be seen in the light of utilising the body of knowledge on existing drugs, rather than as isolated examples of inorganic chemistry. To do this, it is important to have as great as possible an understanding and appreciation of the myriad factors which determine the success or otherwise of a clinical agent. The chapters here have been chosen for this purpose as they all strive to combine the chemical understanding of inorganic molecules with their biological activity. It is the purpose of this monograph to present overviews of the chemistry and biology both of well-known clinically used agents and of new potential applications. Rather than attempt an exhaustive listing of all metal-based drugs, we have selected case studies for the reader to study. The introduction in Chapter 1 gives an overview and discusses the outline of the book and will not be repeated here.

The field of drug design and discovery is, by definition, one of the most interdisciplinary. While emphasis tends to be placed on mechanistic studies (target interactions) the pharmacokinetics and tissue distribution are equally important in determining whether a drug will have a sufficient therapeutic index to be of clinical use. Understanding of how chemical structure determines the above factors is a critical cornerstone for the future. Thus, descriptions in this book interweave chemistry with pharmacology and molecular biology. Further, it is important to know and understand how the many structural features of inorganic compounds capable of subtle fine-tuning such as coordination number and geometry, types of ligands and kinetic reactivity may be tailored to the use of medicinal chemistry. In the last years advances have been made on several fronts. More detailed understanding of the current drugs continues to be accumulated, opening new leads for drug design. Advances in genetic and structural biology underscore the genetic basis of many diseases as residing in metalloproteins. These findings offer a rare opportunity for the inorganic chemist to contribute to the genetic understanding of such diseases. Likewise, the trend, especially in cancer, away from an approach of simple testing to drug design based on specific protein and DNA targets also offers an opportunity for the inorganic chemist. What are the design features of a metal-based agent, for example, which would be suitable for current much-studied targets such as topisomerase and telomerase, to name but two? It is our hope that reading these reviews will excite the reader and especially young researchers to think in similar directions, so that future examples of metal-based drugs will continue to appear.

This volume was long in gestation. I thank the authors for their continued enthusiasm and patience. I thank Dorothy Silvers for her wonderful editing and finally my family for their support.

Nicholas P. Farrell

Contents

CHAPTER 1
Overview

NICHOLAS P. FARRELL

Department of Chemistry, Virginia Commonwealth University,
1001 W. Main Street, Richmond, VA 23284, USA

1 Introduction

The field of inorganic chemistry in medicine may usefully be divided into two
main categories – drugs which target metal ions in some form, whether free or
protein-bound, and secondly, metal-based drugs where the central metal ion is
usually the key feature of the mechanism of action. Metal-based drugs are a
commercially important sector of the pharmaceutical business. Applications
continue to grow and approaches to further clinically useful agents are ever more
sophisticated. How to approach this field from a didactic and systematic manner,
rather than a simple listing of clinical and potential uses, is a challenge. Nevertheless, it is important to attempt to do so to harness the diversity of inorganic
chemistry to systematic developments in medicine.

Any consideration therefore of the uses of inorganic chemistry in medicine
must bridge at least two areas – bioinorganic chemistry and medicinal chemistry.
Bioinorganic chemistry is best considered as understanding all aspects of the role
of metal ions in biology and has been traditionally heavily involved in understanding their processing, incorporation into protein and the nature and function of metalloproteins. In a 'steady-state' environment all essential metals are
incorporated in the right place at the right time and the organism functions
normally. Alternatively, genetic factors may lead to failure to incorporate and
subsequent metabolic disorders may be caused by free metal ions.[1] Advances in
our understanding of how cells process metals and the genetic basis of disease is
naturally expanding the traditional directions of bioinorganic chemistry toward
an appreciation of its medical importance – especially with respect to the role of
metalloproteins in human health and disease. Medicinal chemistry requires
intimate knowledge of the metabolism and stability, as well as target interactions
of the drug. Most mechanistic work is performed in tissue culture or with isolated
proteins, DNA and/or RNA. In tissue culture assays to measure the efficacy of a
potential drug in inhibiting cell growth, the drug is usually in direct contact with
medium throughout the experiment. There is not always a direct extrapolation

to the clinically relevant *in vivo* situation when biodistribution and pharmacokinetics play an increasingly important role in determining drug efficacy.
Many compounds with exciting *in vitro* results have failed to display the same
promise *in vivo*. Nevertheless the mechanistic information of tissue culture
experiments is very useful and, aside from target interactions, may also inform on
approaches to *in vivo* efficacy. Finally, medicinal chemistry distinguishes between drugs acting by a pharmacodynamic mechanism and chemotherapeutic
drugs.[2] In the former case, the drug action must be rapid and essentially
reversible. A patient who submits to an anaesthetic does not expect to be
deprived of feeling forever.[2] Further, a graded response is required to balance
effects – a drug to reverse a stroke must be aware of the severity of that stroke and
concentrations adjusted accordingly. Chemotherapeutic agents on the other
hand involve cell killing, an irreversible process.

In this volume we review aspects of the use of inorganic compounds as drugs
and chemotherapeutic agents. The status with respect to some known drugs is
reviewed as well as introductions to newer drugs of potential clinical significance.
We do not intend to be comprehensive but rather present specific case studies for
reading. In this introduction we give a broad overview of the area from a didactic
point of view. In attempting to do so, four main subdivisions logically present
themselves: (i) uses of chelating agents to sequester specific metal ions or metalloproteins; (ii) inorganic-based drugs acting by a pharmacodynamic mechanism;
(iii) inorganic-based chemotherapeutic agents and (iv) inorganic-based imaging
agents. The reader is referred to the many comprehensive reviews in both
bioinorganic and medicinal chemistry for further reading.

2   Metal Ions in Disease. The Use of Chelating Agents

It is well understood that many metals are essential for the human organism and
endogenous concentrations are tabulated in most bioinorganic chemistry textbooks.[3,4] However, a corollary of this situation is that uncontrolled mobilization
may lead to the presence of excess free metal ion, with subsequent health
problems. The classic examples are those of iron and copper overload. Wilson's
disease is an autosomal disorder of copper accumulation, which untreated is
inevitably fatal. Alloyed to this is the prospect of disease occurring through
adventitious exposure to toxic doses of either essential elements and non-essential
elements such as cadmium, mercury and lead. The treatments for copper and iron
overload are well documented and a list of clinically used chelating agents is found
in most textbooks – typical examples are shown in Figure 1. Their chemistry and
toxicology is also very well documented.[5] A major consideration for the improvement of chelating agents is of course that of metal ion selectivity – few chelating
agents can be stated to be specific for simply one metal ion.

Metalloproteins as Drug Targets

A more recent and related question to the specificity of chelating agents is that of
metalloprotein targets. It is not surprising that many metalloproteins and metal-

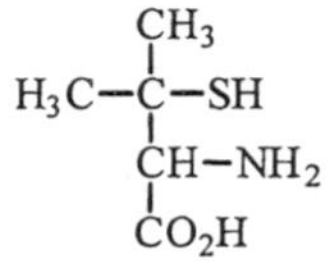

$$
\begin{array}{c}
CH_3 \\
| \\
H_3C-C-SH \\
| \\
CH-NH_2 \\
| \\
CO_2H
\end{array}
\qquad\qquad
\begin{array}{c}
CH_2-SH \\
| \\
CH_2-SH
\end{array}
$$

D- Penicillamine 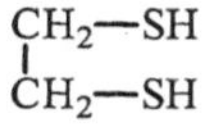   Cysteamine

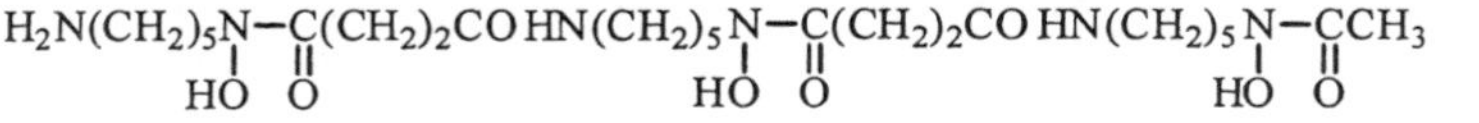

$$H_2N(CH_2)_5 N - C(CH_2)_2CO\ HN(CH_2)_5\ N - C(CH_2)_2CO\ HN(CH_2)_5\ N - CCH_3$$
$$\quad\quad\ \ HO\ \ O \qquad\qquad\qquad HO\ \ O \qquad\qquad\quad HO\ \ O$$

Desferrioxamine B

Figure 1 *Structures of some clinically used chelating agents for treatment of copper and iron overload*

Ioenzymes play vital metabolic roles as well as being critical in genetic information transfer. Drug design and discovery relies more and more on the elucidation of the three-dimensional structure of a target by X-ray crystallography or nuclear magnetic resonance methods, followed by modelling and synthesis of potential inhibitors of the protein or enzyme active site. Metalloproteins are being increasingly recognized and examined as drug targets. Ribonucleotide reductase, the diiron enzyme essential for *de novo* synthesis of deoxyribonucleotides for DNA synthesis has long been recognised as a drug target. The pharmaceutical and chemical properties of chelating thiosemicarbazones and their potential interference with the active site iron moieties has been an extensively studied problem.[6,7]

A current and very relevant example is zinc. Zinc is the second most prominent trace metal in the human body after iron. While deficiency of zinc may cause growth effects, few noxious effects of excess zinc have been observed and zinc *per se* is probably one of the least toxic metals. Zinc is involved in a large number of enzymatic functions, fulfilling both structural and catalytic roles.[8] These functions include DNA transcription and regulation as well as oxidation and hydrolysis, cleavage of peptide bonds as well as formation of phosphodiester bonds. Because zinc is not redox active its catalytic functions derive from its properties as a Lewis acid. More recently, zinc proteins have been recognised as attractive targets for chemotherapy. Especially, there are two principal areas of interest in pharmaceutical laboratories – inhibition of the matrix metalloproteinase enzymes such as collagenase as an approach to treatment of metastatic cancer and inhibition of zinc finger activity as a novel chemotherapeutic attack against HIV infectivity. These apparently diverse goals are united by the common feature that the active-site zinc is in both cases the target of attack.

Zinc and the Human Immunodeficiency Virus

Zinc Fingers as Medicinal Target

The role of zinc in transfer of genetic information is believed to be structural, deriving from the specific conformations proteins adapt upon complexation by the metal. Many transcription factors (required for RNA transcription) contain zinc.[8] The existence of metal-binding domains in regulatory proteins was first postulated because researchers noted that the systematic repeats of cysteine and histidine residues in X_3-Cys-X_{4-4}-Cys-X_{12}-His-X_{3-4}-His-X_4 suggested a role for metal-binding.[9] Model building suggested that zinc binding to His and Cys folded the protein into a conformation, which repeated looked like 'fingers'. X-ray crystallographic evidence has now been obtained for such structures.[10]

The human immunodeficiency virus type 1 (HIV-1) is the etiologic agent of acquired immune deficiency syndrome (AIDS). Effective therapies for AIDS are still urgently required, despite the intense efforts and screening of nearly 220,000 natural and synthetic agents during the last fifteen years. Currently, combination therapy using especially purine and pyrimidine analogs such as ddC, ddI and AZT in conjunction with protease inhibitors is a very promising approach to achieve permanent therapeutic effects. Part of the rationale behind combination therapy has been to use drugs which act on different parts of the viral cycle, thus limiting development of resistance. However, new effective therapies producing long-lasting permanent effects against HIV infectivity are still urgently needed. As such, new targets within the viral cycle need to be identified and understood on a molecular basis to allow development of drug design strategies.

A relatively recent target for drug design has been the zinc fingers of the nucleocapsid protein.[11] A principal approach has been to design chelating agents such as dithiobisbenzamides (DIBAs; Figure 2), which chemically modify the zinc finger cysteine residues resulting in zinc ejection from the fingers with resultant inhibition of HIV replication.[12-14] While these results are highly promising, they are not optimal as yet but serve as a basis for further rational target-based drug design.

Matrix Metalloproteinases (MMPs)

Matrix metalloproteinases are involved in extracellular matrix degradation during cell migration. Normal processes include wound healing, bone remodeling and embryo development. All these functions are highly regulated. The MMP activity is inhibited in normal tissue by endogenous tissue inhibitors of metalloproteinases (TIMPs).[15] Abnormal regulation (elevated peptidase activity) occurs in rheumatoid arthritis and invasion and metastasis of neoplastic cells. The MMPs cleave one or more components of the extracellular matrix – for example collagenase cleaves a specific glycine–leucine bond in collagen.[16] The role of matrix metalloproteinases in metastasis has sparked intense inter-

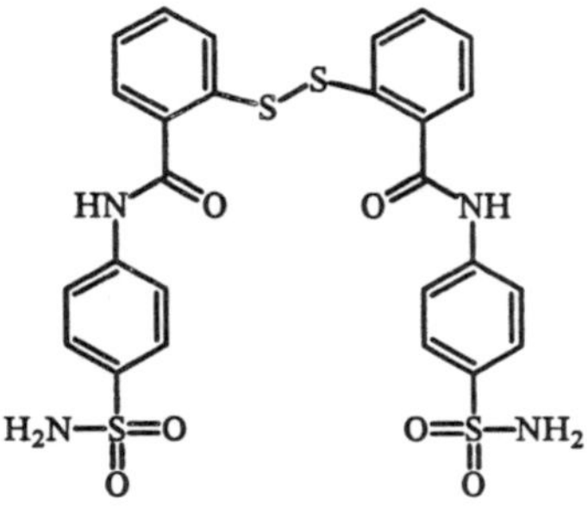

DIBA-1

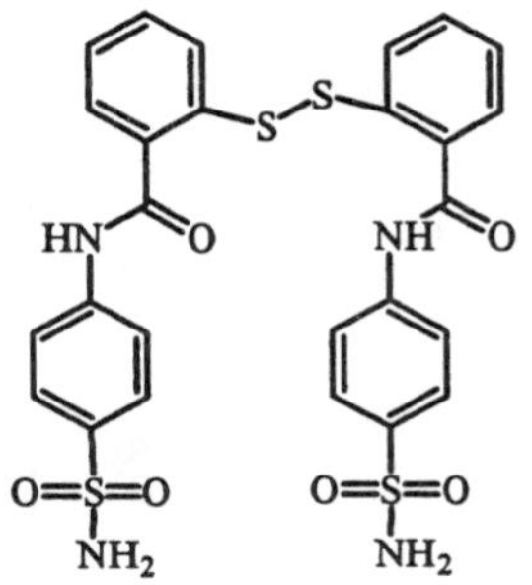

PD022551

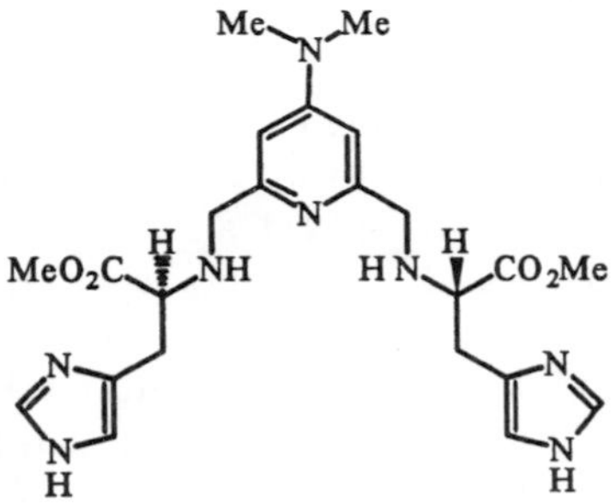

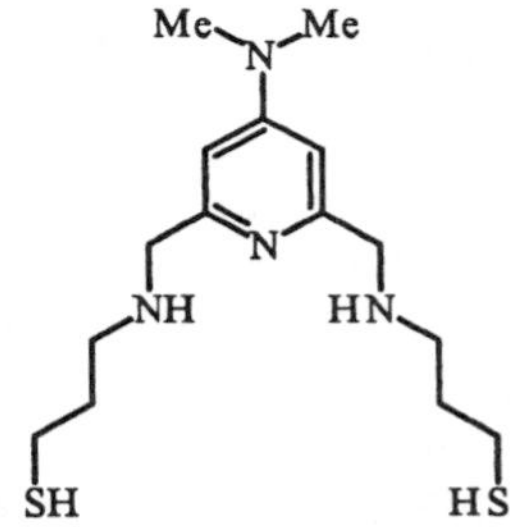

Figure 2 *Structures of chelating agents designed to inhibit Zn-finger function. See references 12–14*

est as a target of chemotherapeutic intervention – very briefly, inhibition of the abnormal regulation of MMPs would in principle retard or eliminate metastasis. The medicinal approach to drug development is to develop as inhibitors substrate analogs which will competitively bind to the zinc active site.[17] The first drug to enter clinical trials from this approach is batimastat (Figure 3). This area is a very active one in drug development and almost all pharmaceutical companies have drug development programs.[18–20] The coming years should see significant advances in drug efficacy and the validation of this approach.

Thus, even in consideration of chelating agents we see that detailed knowledge of metalloprotein structure leads to new approaches to *specific* chelating agents. Indeed a most interesting case is posed by approaches to attacking zinc-finger sites with metal-based compounds such as authiomalate.[21] In this approach the cysteines of the zinc fingers are the specific targets and metal-ligand replacement is a viable goal.

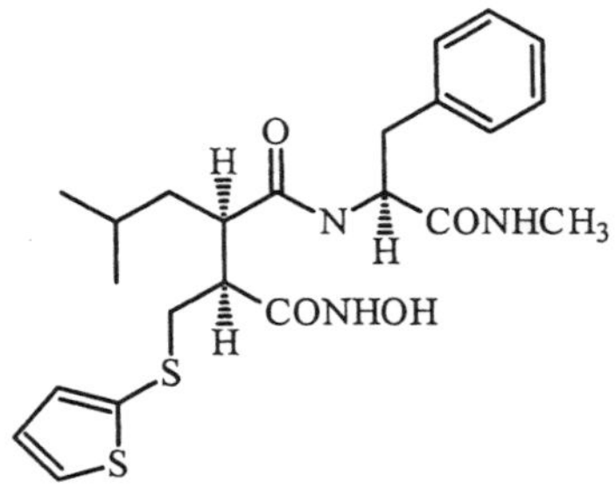

Bastimastat

Figure 3 *Structure of the matrix metalloproteinase inhibitor batimastat*

3 Modulation of Cellular Responses by Metal-containing Drugs

Inorganic drugs may be recognised as acting through a pharmacodynamic mechanism – modulating cellular responses. Clinically used examples discussed in this volume are Li_2CO_3 (Chapter 2) and gold-based antiarthritic drugs (Chapter 3). The further potential for gold-based chemotherapeutic agents is summarized in Dr Shaw's contribution. Since the recognition of the messenger role of the small inorganic molecule NO in the early 1990s, a significant body of data has been accumulated (Chapter 4). From the perspective of inorganic drugs in medicine, the understanding of the importance of NO also made clear how the vasodilating properties of nitroprusside are manifested – through release of NO. Sodium nitroprusside is used in cases of severe heart failure and is a short-acting hypotensive drug with a duration lasting 1–10 minutes. This consideration is an excellent example of the difference between pharmacodynamic and chemotherapeutic drugs. Interesting developments may be expected as inorganic chemists design newer M–NO molecules with appropriate release rates as well as molecules which may scavenge endogenous NO or its potentially damaging oxidation products such as peroxynitrite. Chapters 5 and 6 introduce two exciting and potentially significant advances in the stable of metal-based drugs – manganese-based superoxide dismutase mimics (Chapter 5) and vanadium-based insulin mimetics (Chapter 6).

4 Metal-based Chemotherapeutic Drugs

Chemotherapy is the use of drugs to injure an invading organism without injury to the host. This definition therefore covers the antibacterial, antiviral and anticancer agents. In the first two, the invading organism is clearly distinct from the host. In the case of cancer, a family of diseases characterized by uncontrolled cellular proliferation, the organism is strictly not different but the treatment has a common aim, that of elimination of the unwanted cells. Thus, chemotherapeutic drugs, in contrast to pharmacodynamic drugs, must induce an irreversible cytotoxic effect.

By far the greatest success of inorganic chemotherapy is the advent of cisplatin and carboplatin into the clinic. The current status is outlined in Chapter 7 by Kelland. In the platinum field, the strict reliance on analog development based on the cisplatin structure has produced many promising compounds but, again, few are likely to advance to full clinical use. All direct structural analogs of cisplatin produce a very similar array of adducts on target DNA and it is therefore not surprising that they induce similar biological consequences. This latter consideration led us to formulate the hypothesis that development of platinum compounds structurally dissimilar to cisplatin may, by virtue of formation of different types of Pt–DNA adducts, lead to compounds with a spectrum of clinical activity genuinely complementary to the parent drug. We considered that future discovery of clinically useful platinum agents was likely to arise with 'non-classical' structures. This has been successful to the point that, at time of writing, a novel trinuclear platinum agent has just entered clinical trials. Chapter 8 summarizes the chemical and biological activity of this important new agent.

Will clinical success in the treatment of cancer be limited to platinum-containing drugs? The discovery of the anticancer activity of cisplatin sparked intense interest and research to find other metal-containing anticancer agents. This effort has been well documented and there are now many distinct classes of metal-based drugs with antitumour activity in experimental models.[22–26] Unfortunately, none has as yet achieved full clinical use, let alone the status of cisplatin. Another possibly exciting development is the recognition of certain ruthenium compounds as metastatic poisons rather than cytotoxic agents.[27] Finally the natural product bleomycin is always classified as an inorganic-based drug through the imputed DNA strand breakage mechanism facilitated by oxygen radical production on iron. The chemistry and biology are covered in Chapter 9.

Many common antibacterial agents are silver- and mercury-based – such as silver sulfadiazene and mercurochrome. Their uses and purported mechanisms have been summarized in a previous monograph[22] and will not be presented in great length in this volume.

5   Metal Complexes as Diagnostic Agents

A further subset of inorganic drugs in medicine is comprised of diagnostic agents. In this application, no pharmacodynamic or chemotherapeutic end is desired. Rather, imaging of tissue is achieved. The two principal sets are technetium-based imaging agents and paramagnetic MRI contrast agents. The considerations for clinical development of imaging and contrast agents are somewhat similar to those for drugs – stability and water solubility are paramount. The considerations for stability are important to maintain specific tissue imaging and safety. Clearly, the agent must be relatively unreactive and not be rapidly metabolized or degraded. Finally, clearance of the agent must also be relatively rapid. Clinical experience with both gadolinium[28] and technetium[29] agents is summarized in appropriate pharmceutical reference books.

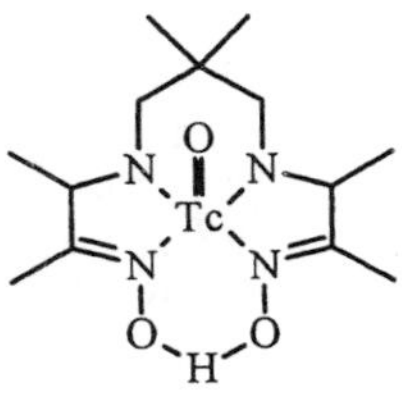

Tc- (HMPAO) 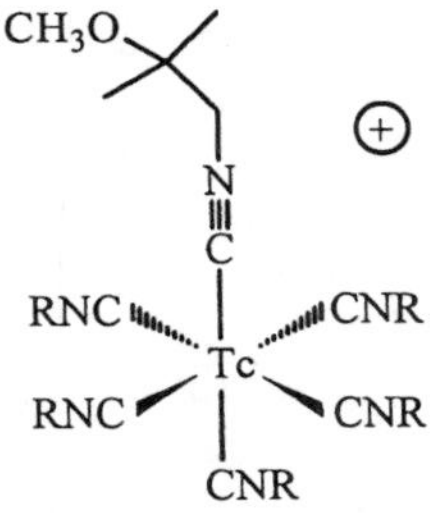   Tc- (MIBI)

Figure 4 *Structures of clinically useful technetium imaging agents*

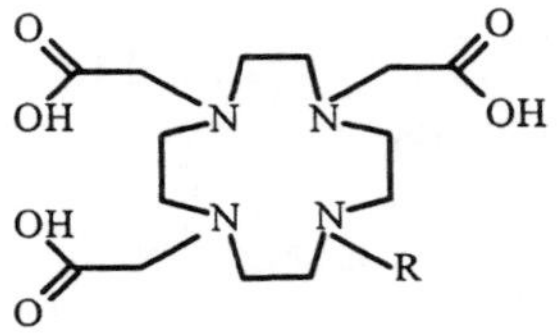

R = H   DO3A
R = CH2CH(OH)CH3   HP-DO3A
R = CH2COOH   DOTA

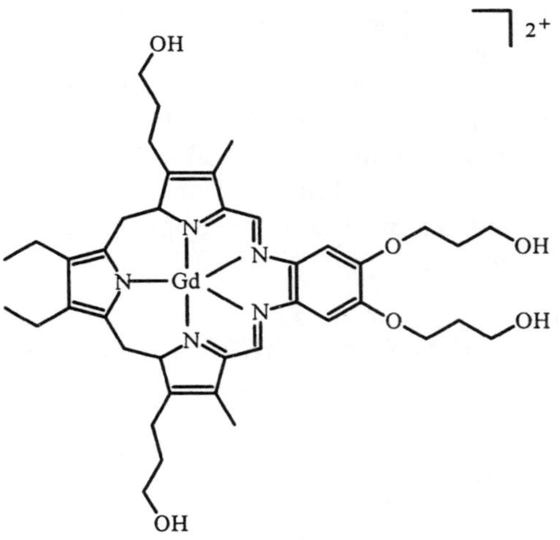

Figure 5 *Structures of Gd-based MRI contrast agents*

Selected examples of clinically used technetium compounds are shown in Figure 4. The challenges of further tissue specificity is an active one while the technology to deliver and prepare coordination complexes with the radioactive isotope is advanced.[30] Likewise, the use of gadolinium agents in MRI work is now well established and the 'classic' chelate structures first developed are shown in Figure 5.[31,32] Future generations of agents may include larger chelates such as the 'texaphrins', an example of which is also shown in Figure 5.[33] Finally, the *in vivo* use of metal radioisotopes in cancer detection and imaging is an important and well documented use of metal chelates.[34]

6 Summary and Acknowledgements

This necessarily brief overview shows the wide scope of inorganic chemistry in medicine. Many examples are by now well documented but the field remains an active one for research and improvement. Besides the principal references given here, many aspects of this subject are covered in specific volumes such as those of *Metal Ions in Biology*. Much useful information can also be found in encyclopedic handbooks.[35] Finally it is a pleasure for me to acknowledge the assistance and understanding of all the contributors and of The Royal Society of Chemistry.

References

1 T.V. O'Halloran, *Science*, 1993, **261**, 715.
2 A. Albert, *Selective Toxicity*, 6th Edn., Wiley and Sons, New York, 1979.
3 J.A. Cowan, *Inorganic Biochemistry. An Introduction*, VCH, New York, 1993.
4 J.J.R.F. da Silva and R.J.P. Williams, *The Biological Chemistry of the Elements*, Clarendon Press, Oxford, 1991.
5 *Chelating Agents Antidotes and Antagonists*, in *Martindale. The Extra Pharmacopoeia*, 31st Edn., Royal Pharmaceutical Society, London, 1996, p. 973.
6 J.G. Cory, G. Rappa, A. Lorico, M-.C. Liu, T.-S. Lin and A.C. Sartorelli, *Advan. Enzyme Regul.*, 1995, **35**, 55.
7 E.C. Moore and A.C. Sartorelli, in *Inhibitors of Ribonucleotide Diphosphate Reductase Activity, International Encyclopedia of Pharmacology and Therapeutics*, Section 28, eds. J.G. Cory and A.H. Cory, Pergamon Press, New York, 1989, p. 203.
8 B.L. Vallee and D.S. Auld, in *Interface between Chemistry and Biology*, eds. P. Jolles and H. Jornvall, Verlag, Basel, 1995, p. 259.
9 A. Klug and J.W. Schwabe, *FASEB J.*, 1995, **9**, 597.
10 N.V. Pavlevitch and C.O. Pabo, *Science*, 1991, **252**, 809.
11 R.N. de Guzman, C.C. Stalling, L. Pappalardo, P.N. Borer and M.F. Summers, *Science*, 1998, **279**, 384.
12 W.G. Rice, C.A. Schaeffer, L. Graham, D. Clanton, R.W. Buckheit Jr., D. Zaharevitz, M.F. Summers, A. Wallqvist and D.G. Covell *J. Med. Chem.*, 1996, **39**, 3606.
13 P.J. Tummino, J.D. Scholten, P.J. Harvey, T.P. Holler, L. Maloney, R. Gogliotti, J. Domagala and D. Hupe, *Proc. Natl. Acad. Sci. USA*, 1996, **93**, 969.
14 M. Otsuka, M. Fujita and Y. Sugiura, *J. Med. Chem.*, 1994, **37**, 4267.
15 F.-X. Gomis-Ruth, K.M., M. Betz, A. Bergner, R. Huber, K. Suzuki, N. Yoshida, H. Nagase, K. Brew, G.P. Bourenkov, H. Bartunik and W. Bode, *Nature*, 1997, **389**, 77.
16 E. Baramova and J.-M. Foidart, *Cell Biol.y Int.*, 1995, **19**, 239.

17 P.D. Brown, *Adv. Enzyme Regul.*, 1995, **35**, 293.

18 B. Lovejoy, A.M. Hassell, K. Longley, M.A. Luther, D. Weigl, G. McGeehan, A.B. McElroy, D. Drewry, M.H. Lambert and S.R. Jordan, *Science*, 1994, **263**, 375.

19 F.K. Brown, D.M. Bickett, C.L. Chambers, H.G. Davies, D.N. Deaton, D. Drewry, M. Foley, A.B. McElroy, M. Gregson, G.M. McGeehan, P.L. Myers, D. Norton, J.M. Salovich, F.J. Schoenen and P. Ward, *J. Med. Chem.*, 1994, **37**, 674.

20 K. Darlak, M.S Stack, A.F. Spatola and R.D. Gray *J. Biol. Chem.*, 1990, **265**, 5199.

21 M.L. Handel, A. deFazio, R.O. Day and R.L. Sutherland, *Proc. Natl. Acad. Sci. USA*, 1995, **92**, 4497.

22 N. Farrell, *Transition Metal Complexes as Drugs and Chemotherapeutic Agents*, in *Catalysis by Metal Complexes*, eds. B.R. James and R.Ugo, Reidel-Kluwer Academic Press, Dordrecht, 1989, Vol. 11.

23 S.P. Fricker, *Metal Compounds in Cancer Therapy*, Chapman and Hall, London, 1994, Vol. 1.

24 *Metal Complexes in Cancer Chemotherapy*, ed. B. Keppler, VCH, Basel, 1993.

25 I. Haiduc and C. Silvestru, *Organometallics in Cancer Chemotherapy*, CRC Press, Boca Raton, 1990, Vol. 2.

26 *Progress in Clinical Biochemistry and Medicine*, Vol. 10, 1989.

27 G. Sava, S. Pacor and S. Zorzet, *Pharmacology (Life Science Advances)*, 1990, **9**, 79.

28 *Contrast Media*, in *Martindale. The Extra Pharmacopoeia*, 31st Edn., Royal Pharmaceutical Society, London, 1996, p. 1009.

29 *Radiopharmaceuticals*, in *Martindale. The Extra Pharmacopoeia*, 31st Edn., Royal Pharmaceutical Society, London, 1996, p. 1463.

30 M.J. Clarke and L. Podbielski, *Coord. Chem. Rev.*, 1987, **78**, 253.

31 P. Armstrong and S.F. Keevil, *Br. Med. J.*, 1991, **303**, 105.

32 R. Lauffer, *Chem. Rev.* 1987, **87**, 901.

33 J. Sessler *et al.*, *J. Am. Chem. Soc.*, 1993, **115**, 10368.

34 D.J. Hnatowich, in *Metal Compounds in Cancer Therapy*, ed. S.P. Fricker, Chapman and Hall, London, 1994, Vol. 1, p. 215.

35 *Handbook of Metal-Ligand Interactions in Biological Fluids*, ed. G. Berthon, Marcel-Dekker Inc., New York, 1995, Vols. 1 and 2.

CHAPTER 2

Biomedical Uses of Lithium

NICHOLAS J. BIRCH

Academic Consultancy Services Limited, Codsall, Staffordshire WV8 2ER, UK

1 Introduction

Lithium is a chemical and biological oddity. Its chemistry is perceived as simple and relatively uninteresting except for its organo-metallic compounds; its biological effects are not easy to relate to specific loci of action. The fascination of lithium, however, is that it is the lightest and smallest solid element, with apparently little subtle chemistry to its credit, and yet it has major and diverse effects in the body. Recent studies of the biological effects of lithium have stretched the scope of its potential uses to include treatments for major disease types such as neoplasms, retroviral infections, and immunological disorders, and for this reason we must make parallel advances in the chemical and biological understanding of its cellular mode of action. Since the tiny lithium ion can cause such significant changes in biological systems, it is clear that, whatever the disease process involved, there is strong possibility that studies at the molecular level will shed some light on the disease itself. The first comprehensive pharmacological review of lithium was carried out by Mogens Schou in 1957.[1]

2 Background

Lithium extraction takes place mainly in North and South America from rocks and brines associated with volcanic activity and aridity. Most of the world's lithium is used in the production of lightweight metal alloys, glass, lubrication greases, and electrical batteries. Only a small proportion of lithium production (less than 1% of the total) is used in medicine.

Lithium occurs naturally in biological tissues and hence is incorporated into foodstuffs.[2] It occurs widely in drinking water, usually at low concentrations. Natural waters that contain higher concentrations of this and other metals frequently are designated 'mineral waters', with supposed medicinal properties. Lithium was first used medically for the treatment of gout. Garrod (1859) first described its medical use in detail and particularly mentioned the use in 'brain gout', a depressive disorder.[3] Lithium urate is the most soluble salt of uric acid and hence was expected to increase uric acid excretion to relieve gout symptoms.

Lithium's clinical value in psychiatry was discovered in 1949 by Cade, an Australian psychiatrist.[4] At the time there were no effective treatments for any of the major psychiatric diseases, and the observation of the effect of lithium must therefore have been startling and exciting. Since the mid-1960s, lithium's use has escalated until it is now estimated that about 500,000 patients receive it, worldwide. The lack of potential for major commercial exploitation has restricted the development of lithium as compared with organic psychotherapeutic agents. Nevertheless, it is used by 60,000 patients in the United Kingdom alone, with an annual £25M saving to the health service.

Lithium carbonate is used specifically for the prophylaxis or prevention of recurrent mood changes in patients suffering from manic depressive psychoses, the recurrent affective disorders.[5] It is of limited use for other psychiatric states, with the possible exception of pathological aggression, where it does seem to have a role to play. Despite many scares, lithium is a very safe drug in experienced hands. The ability of lithium to reduce or abolish recurrent mood swings has undoubtedly improved immensely the quality of life of many patients and their families and saved the lives of many who would otherwise have been led to suicide.[6,7]

Lithium is administered orally, usually as lithium carbonate in tablet form at a total dose of up to 30 millimoles (2 g) per day. Treatment is monitored using regular estimations of blood lithium taken twelve hours after the previous dose.[5,8]

Determination has traditionally been by Flame Emission Spectrometry or Atomic Absorption Spectroscopy, but lithium-ion-selective electrodes are now commercially available and are the most effective way of ensuring compliance in lithium-treated patients.[9] Not only are they accurate within the clinically useful range, but they also provide the long-sought opportunity for the psychiatrist to measure blood lithium in the presence of the patient, who can thereby be challenged about non-compliance or counselled, and the dose regulated accordingly. The usual delays in laboratory reporting of results are avoided, so that the monitoring and dose adjustment or counselling is simultaneous, enhancing the patient's perception of the treatment. Serum lithium concentrations should lie in the range 0.4–0.8 mmol l^{-1}. Higher levels may be associated with side-effects, including tremor, dizziness, drowsiness, and diarrhoea.[8,10] Mood stability may not occur until after a significant period on lithium medication; the time lag may be in the order of months rather than weeks.[11]

Minor side effects may be experienced. They generally occur within four hours of the dose, when serum lithium concentrations are at their highest. Side effects can be minimised by operating with a lower therapeutic range. The most common side effect seen on initiation of therapy is a fine hand tremor, which disappears after a few weeks of treatment. Transient, non-toxic side effects such as dry mouth and nausea also may be experienced by patients when serum lithium concentrations are within the therapeutic range.[5,11] After prolonged therapy, a number of other side-effects have been noted, including mild leucocytosis, hypothyroidism, weight gain, and hypoparathyroidism. Renal and thyroid function should be assessed at least once every year.[11]

Serious intoxication is rare except when therapy is not well controlled, though it is very occasionally seen as a consequence of an unsuccessful suicide attempt.[12] In a study of lithium intoxications in a group of patients whose total exposure time was about 4900 patient years, 15 cases of deliberate self-poisoning were seen with no fatalities.[13,14]

Disturbing reports in the late 1970s linked long-term lithium therapy with renal damage and polyuria. These findings have now been discounted, following an extensive range of studies carried out in major laboratories throughout the world.[15-19] Cohort studies have confirmed that in practice there is a low incidence of renal disease amongst the lithium groups.[12,20]

3   New Uses of Lithium

During the past decade, interest in lithium has entered a new phase, with the discovery of effects unrelated to its psychiatric use. Many of these effects derive from the well-established modification by lithium of hematopoietic processes, notably its stimulation of leucocytosis, which occurs due to enhanced myelopoiesis and to alterations in the marginated pool of polymorphonuclear leucocytes. Colony-stimulating factors from bone marrow macrophages are increased in the presence of lithium. Initially this effect was exploited to treat drug-induced hematopoietic suppression,[21,22] for example, in chemotherapy of cancers[23,24] following bone marrow transplantation,[22] or in radiation-induced injury.[25,26] The metal also affects the immunologic responses to a number of challenges.[27] In the process of these investigations, however, it has become clear that lithium can influence a series of cytokines that regulate such cell differentiation not only in blood-forming cells, but also in other cell types.[28-30]

Lithium influences many aspects of blood cell production,[31] in particular the formation of granulocytes.[32,33] Lithium is effective whenever granulocyte production is either faulty or inadequate.[22,34] The anti-viral drug Zidovudine (AZT) has been used extensively in the treatment of acquired immune deficiency syndrome (AIDS), but its effectiveness is limited by the myelosuppression and bone marrow toxicity that it induces.[35,36] A model of human AIDS is found in mice infected with the Murine Acquired Immune Deficiency Syndrome virus (MAIDS). Lithium, when joined with AZT and combined *in vitro* with normal bone marrow cells, or when administered *in vivo* to mice receiving dose-escalation AZT, significantly reduces the myelosuppression and marrow toxicity of AZT.[21,36] Further studies have demonstrated lithium's capacity to modulate AZT toxicity by influencing blood cell production, when administered to normal mice following an initial exposure to AZT.[37,38] Animals receiving AZT alone showed anaemia, thrombocytopaenia, and neutropaenia, which were dose-related and which were prevented by combination lithium/AZT treatment. These studies support the view that lithium might play a part in the treatment of HIV-infected patients receiving anti-viral therapy. Lithium also appears to reverse the lymphoma associated with MAIDS infection, which is analogous to that seen in human AIDS.[39-42]

Several processes in the immune response are affected by lithium *in vivo* and *in vitro*.[27,32,43-45] The proliferative responses of hamster lymphoid cells to concanavalin A or phytohaemagglutinin were enhanced by lithium in a serum free culture system. Parietal cell, thyroid, and anti-nuclear antibodies and pituitary auto-antibodies have been reported to increase in lithium-treated patients.[46] Tissue-specific auto-antibodies may or may not be related to autoimmune disease: lithium, in any case, may modulate pre-existing autoimmune processes rather than induce them. Lithium induces release of enzymes from human granulocyte granules *in vitro*. Lithium also potentiates enzyme secretion by receptor-independent cytochalasin, but the secretory response to receptor-dependent chemotactic peptide f-MLP was not increased. Lithium in high concentration ($10 \, mmol \, l^{-1}$) induced release of lactoferrin from human granulocytes *in vitro*, but lithium effects on other aspects of granulocyte function are less clear; there are contradictory reports of its action.

The lithium salt of γ-linolenic acid (LiGLA) is being assessed for its effects in reducing the progression of a range of cancers, following the demonstration *in vitro* of cytotoxicity of γ-linolenic acid and its salts towards a range of malignant cells.[47-50] LiGLA also is cytotoxic towards AIDS-infected cells, a finding that may have further significance for the treatment of AIDS.[51] Lithium itself appears to have some anti-tumour effect:[24,41] it stimulates the release of Tumour Necrosis Factor (TNF) from stimulated microphages.[52,53] It may be that the cytotoxicity of the GLA salt is a combined effect of the essential fatty acid and the lithium ion.

Lithium is used as a topical application to treat skin diseases.[54] Lithium succinate ointment has proved to be useful and is now licensed for use in the treatment of seborrhoeic dermatitis, having an effect both on the lipid metabolism of the normal skin fungus, *Pityrosporum ovale*, excessively proliferated in this condition, and on the general inflammation that is the normal response to such fungal attack.[55] High concentrations of lithium, about $40 \, mmol \, l^{-1}$, inhibit the replication of herpes, pox, and adenovirus (DNA) viruses, but not of RNA viruses such as influenza encephalomyocarditis.[56] A double-blind, placebo-controlled trial of ointment containing 8% lithium succinate showed that more rapid healing of herpetic ulcers occurred and viral excretion was reduced. There also was decreased duration of pain in patients with recurrent genital herpes simplex infection.[56] Further antiviral and topical uses are expected, as the area is rapidly developing.

Lithium has also been used in the measurement of cardiac output, where it can act as a suitable non-toxic marker that is easily detected by an ion-selective electrode and that, in the time scale of the measurement (30 seconds), is not taken up into cells.[57]

4   Chemistry of Lithium

Alkali metals (Group 1A) easily lose an electron to yield a univalent cation. Alkali metal compounds are almost entirely ionic, although lithium has rather more tendency to form covalent compounds. In solution, the very small diameter of lithium in relation to the aqueous solvent results in a large hydration sphere of

Table 1 *Properties of Group I and Group II metals*

	Li	Na	K	Mg	Ca
Atomic radius (Å)	1.33	1.57	2.03	1.36	1.74
Ionic radius (Å)	0.60	0.95	1.33	0.65	0.99
Hydrated radius (Å)	3.40	2.76	2.32	4.67	3.21
Polarizing power z/r^2	2.80	1.12	0.56	4.70	2.05
Electronegativity	1.0	0.9	0.8	1.2	1.0

Key: Å = Angstrom unit = 10^{-10} m. z = charge on ion (integer). r = radius of ion (Å).

uncertain size (Table 1). In an aqueous environment, the radius increases out of proportion to the radii of the other Group 1A elements, resulting in poor ionic mobility and low lipid solubility under physiological conditions. In other words, in solution the small diameter of lithium results in non-conformity to ideal solution behaviour.[58]

Lithium is the lightest solid element: it has the smallest ionic radius of the alkali metals and the largest field density at its surface. Moreover, lithium is the least reactive of the alkali metals. The atomic and ionic radii of lithium and magnesium are similar; the electronegativity is the same as for calcium. The hydrated radius and polarising power of lithium lie between those of magnesium and calcium. The chemistry of lithium classically is described in relation to that of magnesium by the so-called 'diagonal relationship', and lithium may interact with magnesium and calcium-dependent processes in physiology.[59,60]

5   Distribution of Lithium in the Body and in Cells

Early studies showed that lithium is widely distributed in tissues following oral administration, or intraperitoneal or intravenous injection to experimental animals.[61–72] These reports established the basic understanding of the distribution and rate of uptake of lithium in a number of tissues. Later work demonstrated accumulation in bone[66,68,73,74] and endocrine glands.[75] Thellier, in a series of elegant neutron activation experiments using the isotope ^{6}Li, provided visual localisation of lithium in the whole body[76] and in the different areas of the brain.[77] It is clear that, of the soft tissues, the pituitary, thymus, and most particularly the thyroid glands have rather more than average accumulations of lithium. Bone also appears to accumulate lithium to a limited degree (Figure 1).

While it is clear that no tissue appears to have a very high lithium content, high concentrations may occur locally. At a plasma concentration of 1 mmol l^{-1}, it is possible that the concentration of lithium in the renal papilla during relative water deprivation may reach a concentration of some 60–65 mmol l^{-1}, as may be seen in urine samples from water-deprived subjects.[75] Similarly, local concentrations in the fluids bathing cells of the gastrointestinal tract during absorption of a single lithium tablet may be high. Most cells of the body, however, are exposed to external lithium concentrations of less than 2 mmol l^{-1} even at the highest lithium doses.

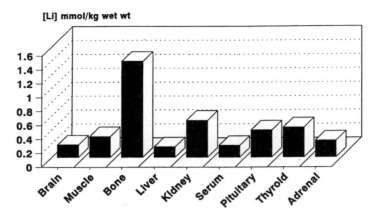

Figure 1 *The distribution of lithium in a range of tissues following chronic administration in rats*

By contrast, in other studies, animals have been maintained from weaning onwards specifically on either low lithium diets (0.005–0.015 ppm) or control diets having 'normal' lithium content. Patt *et al.* followed the experimental groups through three generations of progeny;[78] analysis was by Flame Atomic Absorption Spectrometry using a high temperature, nitrous oxide–acetylene flame. The distributions revealed were generally in line with the findings of lithium administration experiments. However, lithium accumulated significantly in bone and teeth, and to a lesser extent in the anterior pituitary and adrenal glands.

The most significant finding of this study[78] was the contrast between the pituitary and adrenal glands, on the one hand, and all other tissues. The lithium concentration in the two endocrine organs was maintained through three generations despite dietary restriction of lithium, while in all other organs lithium content continued to decline with prolonged deprivation. The fertility of second- and third-generation female rats in the lithium-deprived groups was reduced though there was no apparent effect on the growth rate of the young. Eichner and Opitz[79] demonstrated a sex difference in 'normal' lithium content of both adrenals and thymus glands, adding to the evidence for a physiological role of lithium in endocrine function.

6   Studies Using Lithium Isotopes

Tissue Localisation and Transport of Lithium

One of the problems in the study of lithium action is the lack of precision in localisation of the ion and in the measurement of its movements between cells and between tissues. This lack of precision arises partly because lithium is a very mobile ion, partly because of its widespread distribution in the body, and partly because of the difficulties of lithium analysis. Analytical problems generally stem not from lack of sensitivity in the analysis, but from the interference of related metals and common anions present in large quantities in animal tissues. Thellier and Wissocq[80] have described in detail a number of useful analytical techniques for the biological study of lithium.

Lithium metabolism and transport cannot be studied directly, because the lack of useful radioisotopes[81] has limited the metabolic information available. Lithium has five isotopes, three of which have extremely short half lives (0.8, 0.2, 10^{-21} s). Lithium occurs naturally as a mixture of the two stable isotopes ^{7}Li (95.58%) and ^{6}Li (7.42%), which may be determined using Atomic Absorption Spectroscopy,[82] Nuclear Magnetic Resonance Spectroscopy,[83] or Neutron Activation analysis.[77,80] Under normal circumstances it is impossible to identify isotopes by using AAS, because the spectral resolution of the spectrometer is inadequate. We have previously reported the use of ISAAS in the determination of lithium pharmacokinetics.[82] Briefly, the shift in the spectrum from ^{6}Li to ^{7}Li is 0.015 nm which is identical to the separation of the two lines of the spectrum. Thus, the spectrum of natural lithium is a triplet. By measuring the light absorbed from hollow cathode lamps of each lithium isotope, a series of calibration curves is constructed, and the proportion of each isotope in the sample is determined by solution of the appropriate exponential equation. By using a dual-channel atomic absorption spectrometer, the two isotopes may be determined simultaneously.[82–84]

The microlocalisation technique with the stable isotope ^{6}Li uses a beam of neutrons in an atomic reactor. The ^{6}Li nucleus absorbs a neutron and immediately undergoes fission to produce an α-particle and a ^{3}H atom, which create tracks in a suitable detector placed in contact with ^{6}Li-containing tissue. The tissue distribution in the rat,[85] brain lithium distribution in the mouse[86–89] and the rat,[90,91] distribution in the mouse embryo,[76] kinetics in the mouse brain,[92] and distribution in mutant strains of mice with dysmyelination[93] have been studied.

We have carried out extensive studies on the intestinal transport of lithium and magnesium in human and animal tissue,[94,95] which suggest that these metals are absorbed in the intestinal tract via a paracellular route, that is, *via* the pericellular spaces, which are regulated by the epithelial tight junctions. These metals pass not through, but instead around the intestinal cell in their passage into the body from the gut.[94,96,97] This makes good biological sense, since otherwise intestinal cells would be required to regulate the intracellular concentration of a wide range of toxic and non-toxic metals, some essential, some not, in order to protect the *milieu interieur* of the intestinal cells themselves. We have shown that active processes occur to a small extent in the absorption of magnesium,[98] but that these are likely to be involved in the defence of the cell against magnesium deficiency rather than as a way of regulating gross cell content of the element.[95] It is likely that such processes occur during the import of many metals into the body from the intestinal lumen. It should be recognised that where transport proteins have been identified for particular metals, it is probable that these are for the protection of individual cells rather than a mechanism for bulk uptake of metals into the body.[99]

Cellular Localisation of Lithium

Using ^{7}Li NMR, it is possible to differentiate between atoms or ions that are within the cell and those that are free in the extracellular bathing fluid.[83] We have recorded 31 MHz ^{7}Li NMR spectra, using a Bruker WP80SY multinuclear spectrometer. Cells were incubated in phosphate-buffered saline containing 20% deuterium oxide (as heteronuclear lock) and up to $5 \, mmol \, l^{-1}$ dysprosium tripolyphosphate (shift reagent).

Before carrying out studies on three different types of living cells, we were able to confirm in general terms that in the presence of a dysprosium shift reagent, ^{7}Li NMR signals were detectable from the interior of phosphatidylcholine (PC) liposomes, and that these signals disappeared following addition of the ionophore gramicidin D.[83] We have carried out extensive studies in human and animal erythrocytes (red blood cells) and less extensive experiments in cells obtained from animal sources.[83,100,101] Both ^{7}Li NMR and ISAAS studies were carried out on erythrocytes loaded by incubation in a range of lithium buffers. These experiments show that in previously untreated subjects the erythrocyte internal lithium concentration is under 8% of the external after incubation in a range of external lithium concentrations between 2 and $40 \, mmol \, l^{-1}$ for up to three hours.[102,103] The results from NMR and ISAAS methods are in close agreement. Other workers have reported broadly comparable results,[104–106] including studies in which an alternative method, Modified Inversion Recovery, was used to determine the intracellular component of the lithium signal.[107]

For all of these cell types, we have come to the conclusion that lithium is less readily transported across the cell membrane than has hitherto been believed. It is clear from studies of erythrocytes, hepatocytes, fibroblasts, and astrocytoma cells[83,100,108,109] that lithium does not distribute at equilibrium according to the cellular membrane potential (in these examples, between -40 and $-60 \, mV$). The explanation may be either low membrane permeability or, alternatively, a mechanism of effective ejection of the ion from the cell interior. The situation may be different in more highly polarized cells and those that are electrically active in the nervous system. A number of ion channels are known to accept lithium relatively easily, and under ideal conditions these cells may show significant lithium currents. However, once again, this would not imply substantial cellular lithium accumulation if there is also an effective lithium extrusion mechanism. The case for significant cellular uptake of lithium, even in electrically active cells, is not sufficiently proven to be taken for granted, and may in fact present a variable that is frequently ignored in developing models for lithium's pharmacological action.

7 Biochemistry of Lithium

Magnesium and Calcium

Magnesium is an activator for more than 300 enzymes and has a critical role in the transfer, storage, and utilisation of energy. Its predominant role is as an activator of phosphate transfer reactions, including the hydrolysis and transfer of organic phosphate groups, particularly reactions involving ATP. Magnesium has a pivotal role in carbohydrate, fat, and protein metabolism.[110,111] If lithium were to compete for sites on these enzymes or on some of the many other magnesium-dependent enzymes, widespread metabolic effects might be expected. It has been suggested that lithium may compete for magnesium and calcium binding sites on biological ligands,[60,112] and indeed a number of studies of the effects of lithium on magnesium-dependent systems have been reported.[113–117]

The physiology of calcium and magnesium is directed towards a number of different functions: structural, signalling, and regulatory. Their cellular activities are controlled by their simple chemistry and osmotic relations, which in turn affect the way in which the body utilises these metals. The regulation of these two elements takes place to some extent reciprocally, although the exact mechanisms of magnesium's regulation have not been fully established.[96,118]

Lithium and the Phosphoinositide Signalling System

Lithium selectively interferes with the inositol lipid cycle,[119,120] which is the basis for the proposal of a unifying hypothesis for lithium actions.[121,122] Lithium reduces the cell concentrations of myoinositol, which would otherwise be converted to phosphatidylinositol; this attenuates the response to external stimuli.[119,123]

Neuronal and hormonal signals may often be transduced *via* receptor-mediated activation of phophoinositidase C (inositol lipid-directed phospholipase C), which converts phosphatidyl-inositol 4,5-bisphosphate (PIP2) to 1,2-diacylglycerol (1,2 DG) and D-inositol 1,4,5-triphosphate [(1,4,5)IP3] in the cell membrane.[124,125] These metabolic products are second messengers: 1,2 DG stimulates protein kinase C and (1,4,5)IP3 releases intracellular calcium from the endoplasmic reticulum. Subsequently, (1,4,5)IP3 is ultimately converted to myoinositol. That in turn is converted to phosphatidylinositol, which is used to replenish PIP2 stores and thus complete the cycle.[119]

Lithium has been shown to inhibit inositol monophosphate phosphatase uncompetitively.[119,126] This is an unusual mode of inhibition, which has catastrophic effects on metabolism.[127] In a metabolic pathway at steady state, the increase in the extent of inhibition is non-linear as the products of earlier enzymes in the cycle begin to accumulate. The primary substrate at an earlier stage along the metabolic path is then rapidly depleted, and the regulation of the system as a whole soon becomes unstable and chaotic.

Lithium also inhibits other enzymes in the interconversion and breakdown of polyphosphoinositides, though not by an uncompetitive mechanism. Either because of the uncompetitive inhibition of the monophosphatase or because of the inhibition of the other enzymes, or, indeed, by a combination of these, lithium reduces the cell concentrations of myo-inositol, which would otherwise be converted into phosphatidylinositol. The reduction in cell inositol content attenuates the brain response to external stimuli.[128] This scheme has been suggested as the mechanism of action of lithium in the affective disorders, since dietary sources of inositol cannot cross the blood–brain barrier, and the brain cell must therefore rely on endogenous supplies, which would become rate-limiting were lithium to act in this manner.[129] It is proposed that the mood disorder, for which lithium is effective in prophylaxis, may be generated by an as-yet-unidentified group of brain cells that are pathologically overactive, and that such overactivity would be preferentially attenuated by this restriction of substrate.[129]

8   Lithium at the Cell Periphery: A Novel Viewpoint

The major theories of lithium action depend upon there being a relatively high concentration of lithium in the cell interior, so it is likely that a radical new approach will be required to interpret the mechanism of lithium's action if our preliminary findings of a low intracellular lithium are confirmed.[130] Modern molecular biology stresses that the cell is regulated from the nucleus. Perhaps we should now reconsider that and rethink our ideas of regulation, making a new assumption that the integrator and regulator of the cell is the pericellular apparatus. The implications of such an hypothesis are very broad. Metals within cells may themselves be a means of cell regulation, which in turn is regulated by metal ions at the cell periphery. Hitherto, research has concentrated on the notion that lithium interferes with intracellular processes. For example, the main theory about lithium action currently in vogue is that it acts *via* phosphoinositide metabolism. The evidence shows that lithium does indeed have effects on phosphoinositide metabolism, but it does not show these effects in psychiatric patients at therapeutic lithium concentrations. Phosphoinositide involvement in the biochemical pathology of manic depressive disorders is far from proven. A pericellular site of action of lithium would provide greater scope for explaining the diversity of the lithium effects that have been reported in the literature over many years.

If we consider that metal ions at the cell surface may be either free or restricted in their movement, this provides us with a new mechanism for signalling and a means for the integration of signals from a variety of sources. By contrast, the intracellular metals may be functionally separate and regulated as part of the *milieu interieur* of the cell. It is possible to envisage a model of cellular regulation that depends on the functioning of the cell membrane as a 'bio-microchip', a flexible and infinitely adaptive analogue of our man-made, rigid, silicon-based, electronic microchips.[130] Charged ions may be densely distributed amongst the protruding, or even embedded, portions of glycoproteins and glycolipids that form the glycocalyx.

The whole surface of the cell could be conceived as a patchwork of such microelectronic 'components' distributed across the surface and at variable depth, each with the potential to be regulated by its own individual molecular neighbourhood of macromolecules and ions, subject to flexible configuration induced by changes in cell morphology and by the docking and undocking of membrane-active biomolecules such as neurotransmitters and second messengers.[130]

9 Conclusion

The use of lithium in medicine is a significant success in the field of inorganic pharmacology and is of particular interest because lithium is the lightest solid element, whose chemistry is relatively simple. It must therefore be assumed that whatever lithium does, its action is on fundamental processes. For this reason it may be important, as a probe, to investigate the molecular interactions between more complex drugs and their receptors. If we can discover what it is that lithium does at a molecular level that makes it so effective in psychiatry, we may gain insights into the most basic features of the cellular response to drugs: lithium does not, after all, have a large and convoluted structure that can make multiple contacts with receptors, which might lead to modification of receptor activation. Whatever lithium does, it achieves because it is a highly charged cation with a large hydrated radius and chemical properties similar to those of magnesium.[130,131]

References

1 M. Schou, *Pharmacol. Rev.*, 1957, **9**, 17.
2 N.J. Birch, *Handbook of Toxicity of Inorganic Compounds*, eds. H. Sigel and H.G. Seiler, Marcel Dekker, New York, 1988, p. 383.
3 A.B. Garrod, *The Nature and Treatment of Gout and Rheumatic Gout*, Walton and Maberly, London, 1859, p. 1.
4 J.F.J. Cade, *Med. J. Aust.*, 1949, 349.
5 M. Schou, *Lithium Treatment of Manic Depressive Illness: A Practical Guide*, Karger, Basel, 1989, p. 1.
6 B. Muller-Oerlinghausen, B. Ahrens, E. Grof, P. Grof, G. Lenz, M. Schou, C. Simhandl, K. Thau, J. Volk, R. Wolf, *et al.*, *Acta Psychiat. Scand.*, 1992, **86**, 218.
7 M. Schou, *Lithium and the Cell: Pharmacology and Biochemistry*, ed. N.J. Birch, Academic Press, London, 1991, p. 1.
8 M. Schou, *Am. J. Psychiatry*, 1989, **146**, 573.
9 N.J. Birch, M.S. Freeman, J.D. Phillips and R.J. Davie, *Lithium*, 1992, **3**, 133.
10 M. Schou, *Side Effects of Drugs Annual: 10*, ed. M.N.G. Dukes, Elsevier, Amsterdam, 1986, p. 27.
11 D.P. Srinivasan and N.J. Birch, *Update*, 1992, **45**, 363.
12 A. Nilsson and R. Axelsson, *Acta Psychiat. Scand.*, 1989, **80**, 221.
13 P. Vestergaard and M. Schou, *Pharmacopsychiatry*, 1989, **22**, 99.
14 M. Schou, H.E. Hansen, K. Thomsen and P. Vestergaard, *Pharmacopsychiatry*, 1989, **22**, 101.
15 D.G. Waller and J.G. Edwards, *Psychol. Med.*, 1989, **19**, 825.
16 O. Hetmar, U.J. Povlsen, J. Ladefoged and T.G. Bolwig, *Br. J. Psychiat.*, 1991, **158**, 53.
17 O. Hetmar, C. Brun, J. Ladefoged, S. Larsen and T.G. Bolwig, *J. Psychiat. Res.*, 1989, **23**, 285.
18 M. Schou, *Encephale*, 1989, **15**, 437.
19 M. Schou, *J. Psychiat. Res,*. 1988, **22**, 287.
20 R.P. Hullin, V.P. Coley, N.J. Birch, D.B. Morgan and T.H. Thomas, *Br. Med. J.*, 1979, 1457.
21 V.S. Gallicchio, N.K. Hughes and K.F. Tse, *J. Internal Med.*, 1993, **233**, 259.

22 V.S. Gallicchio, M.J. Messino, B.C. Hulette and N.K. Hughes. *J. Med. Clin. Exp. Theor.*, 1992, **23**, 195.
23 J.A. Sokoloski, J. Li, A. Nigam and A.C. Sartorelli, *Leukemia Res.*, 1993, **17**, 403.
24 Y. Wu and D. Cai, *Proc. Soc. Exp. Biol. Med.*, 1992, **201**, 284.
25 R.M. Johnke and R.S. Abernathy, *Int. J. Cell Cloning*, 1991, **9**, 78.
26 X.Y. Ke, Y.F. Wang and T.Z. Jia, *Chin. Med. J. Peking*, 1991, **104**, 54.
27 D.A. Hart, *Lithium and Cell Physiology*, eds. R.O. Bach and V.S. Gallicchio, Springer-Verlag, New York, 1990, p. 58.
28 H.E. McGrath, P.M. Wade, V.K. Kister and P.J. Quesenberry, *J. Cellular Physiol.*, 1992, **151**, 76.
29 R. Beyaert, K. Schulze Osthoff, F. Van Roy and W. Fiers, *Cytokine*, 1991, **3**, 284.
30 V.S. Gallicchio, N.K. Hughes, B.C. Hulette and L. Noblitt, *J. Leukocyte Biol.*, 1991, **50**, 580.
31 V.S. Gallicchio, *Lithium and the Blood. Lithium Therapy Monographs*: 4, Karger, Basel, 1991, 150 pp.
32 V.S. Gallicchio, *Lithium and the Blood. Lithium Therapy Monographs*: 4, ed. V.S. Gallicchio, Karger, Basel, 1991, p. 1.
33 V.S. Gallicchio, *Lithium and the Cell: Pharmacology and Biochemistry*, ed. N.J. Birch, Academic Press, London, 1991, p. 183.
34 V.S. Gallicchio and N.K. Hughes, *Lithium*, 1992, **3**, 117.
35 D.S. Israel and K.I. Plaisance, *Clin. Pharm.*, 1991, **10**, 268.
36 V.S. Gallicchio, N.K. Hughes, K. Tse and H. Gaines, *Growth Factors*, 1993, **9**, 177.
37 V.S. Gallicchio, S. Kazmi, E. Townsley, N.K. Hughes, K.F. Tse, K.F.W. Scott and N.J. Birch, *Abstracts of the Tenth International Conference on AIDS, Yokahama, August 1994*, 1994.
38 V.S. Gallicchio, M.L. Cibull, N.K. Hughes, K. Tse, K.W. Scott, N.J. Birch and J. Ling, *Lithium*, 1994, **5**, 223.
39 V.S. Gallicchio, M.L. Cibull, N.K. Hughes and K. Tse, *Pathobiology*, 1993, **61**, 216.
40 V.S. Gallicchio, M.L. Cibull, N.K. Hughes and K.F. Tse, *Lithium in Medicine and Biology*, eds. N.J. Birch, C. Padgham and M.S. Hughes, Marius Press, Carnforth, UK, 1993, p. 181.
41 V.S. Gallicchio, M.L. Cibull, J.K. Morrow, K. Tse and N.K. Hughes, *Magnes. Res*, 1994, **7**, 62 (Abstract).
42 E. Townsley, S. Kazmi, N.K. Hughes, K. Tse, J. Ling, K. Scott, N.J. Birch and V.S. Gallicchio, *J. Trace Microprobe Tech.*, 1995, **13**, 1.
43 V.S. Gallicchio and M.G. Chen, *Blood*, 1980, **55**, 1150.
44 V.S. Gallicchio, *Lithium: Inorganic Pharmacology and Psychiatric Use*, ed. N.J. Birch, I.R.L. Press, Oxford, 1988, p. 93.
45 D.A. Hart, *Lithium and the Cell*, ed. N.J. Birch, Academic Press, London, 1991, p. 289.
46 J.H. Lazarus, *Endocrine and Metabolic Effects of Lithium*, Plenum Medical Press, New York, 1986, p. 1.
47 M.E. Begin, G. Ells, U.N. Das and D.F. Horrobin, *J. Nat. Can. Inst.*, 1986, **77**, 1062.
48 F. Fujiwara, S. Todo and S. Imashuku, *Prostaglandins, Leukotrienes and Medicine*, 1986, **23**, 311.
49 A. Anel, J. Navel and P. Desportes, *Leukaemia*, 1992, **6**, 690.
50 M.E. Begin, U.N. Das and G. Ells, *Prog. Lipid Res.*, 1986, **25**, 573.
51 D. Kinchington, S. Randall, M. Winther and D. Horrobin, *FEBS Lett.*, 1993, **330**, 219.
52 R. Beyaert, K. Heyninck, D. De Valck, F. Boeykens, F. Van Roy and W. Fiers, *J. Immunol.*, 1993, **151**, 291.
53 R. Beyaert and W. Fiers, *Lithium*, 1992, **3**, 1.

54 D.F. Horrobin, *Lithium and the Cell: Pharmacology and Biochemistry*, ed. N.J. Birch, Academic Press, London, 1991, p. 273.
55 D.J. Gould, P.S. Mortimer, C. Proby, M.G. Davis, P.J.W. Kersey, R. Lindskov, A. Oxholm, A.M.M. Strong, E. Hamill, K. Kenicer, C. Green, J.J. Cream, R.J. Clayton, J.D. Wilkinson, A. Davis, B.R. Allen, R. Marks, L. Lever, M.Y. Moss, P.F. Morse, S.I. Wright, D.F. Horrobin and J.C.M. Stewart, *J. Am. Acad. Dermatol.*, 1992, **26**, 452.
56 G.R.B. Skinner, C. Hartley, A. Buchan, L. Harper and P. Gallimore, *Med. Microbiol. Immunol.*, 1980, **168**, 139.
57 R.A.F. Linton, D.M. Band and K.M. Haire, *Br. J. Anaesthesia*, 1993, **71**, 262.
58 K.H. Stern and E.S. Amis, *Chem. Rev.*, 1959, **59**, 1.
59 N.J. Birch, *Br. J. Psychiat.*, 1970, **116**, 461.
60 N.J. Birch, *Nature*, 1976, **264**, 681.
61 V.D. Davenport, *Am. J. Physiol.*, 1950, **163**, 633.
62 J.L. Radomski, H.N. Fuyat, A.A. Nelson and P.K. Smith, *J. Pharmacol.*, 1950, **100**, 429.
63 M. Schou, *Acta Pharmacol. Toxicol.*, 1958, **15**, 115.
64 L. Baer, S. Kassir and R. Fieve, *Psychopharmacologia*, 1970, **17**, 216.
65 H.S. Saratikov, *Dokl. Akad. Nauk*, 1971, **201**, 1255.
66 N.J. Birch and R.P. Hullin, *Life Sciences*, 1972, **11**, 1095.
67 A.S. Saratikov and V. Samoilov, *Zh. Nevropatol. Psikhiat.*, 1972, **71**, 1709.
68 N.J. Birch and F.A. Jenner, *Br. J. Pharm.*, 1973, **47**, 586.
69 M.S. Ebadi, V.J. Simmons, M.J. Hendrickson and P.S. Lacy, *Eur. J. Pharmacol.*, 1974, **27**, 324.
70 J. MacLeod, R.C. Swan and G.A. Aitken, *Am. J. Physiol.*, 1949, **157**, 177.
71 J. Foulks, G.H. Mudge and A. Gilman, *Am. J. Physiol.*, 1952, **168**, 642.
72 P.J. Talso and R.W. Clarke, *Am. J. Physiol.*, 1951, **166**, 202.
73 N.J. Birch, *Clin. Sci.*, 1974, **46**, 409.
74 N.J. Birch, A. Horsman and R.P. Hullin, *Neuropsychobiology*, 1982, **8**, 86.
75 N.J. Birch, *Metal Ions in Biological Systems. Vol. 14*, ed. H. Sigel, Marcel Dekker, New York, 1982, p. 257.
76 J. Wissocq, E. Hennequin, C. Heurteaux, F. Marin, J. Signoret, S. Mebarki and M. Thellier, *Spurenelement Symposium* (4), eds. M. Anke, W. Baumann, H. Braunlich and C. Bruckner, Karl-Marx Universität and Friedrich-Schiller Universität, Leipzig and Jena, 1983, p. 127.
77 M. Thellier, J. Wissocq and C. Heurteaux, *Nature*, 1980, **283**, 299.
78 E.L. Patt, E.E. Pickett and B.L. O'Dell, *Bioinorg. Chem.*, 1978, **9**, 299.
79 D. Eichner and K. Opitz, *Histochemistry*, 1974, **42**, 295.
80 M. Thellier and J. Wissocq, *Lithium and the Cell: Pharmacology and Biochemistry*, ed. N.J. Birch, Academic Press, London, 1991, p. 59.
81 N.J. Birch and J.D. Phillips, *Adv. Inorg. Chem.*, Vol. 36, ed. A.G. Sykes, Academic Press, San Diego, 1991, p. 49.
82 N.J. Birch, D. Robinson, R.A. Inie and R.P. Hullin, *J. Pharm. Pharmacol.*, 1978, **30**, 683.
83 M.S. Hughes, *Lithium and the Cell*, ed. N.J. Birch, Academic Press, London, 1991, p. 175.
84 M.S. Hughes and N.J. Birch, *C. R. Acad. Sci.*, Ser. *III*, 1992, **314**, 153.
85 S.C. Nelson, M.M. Herman, K.G. Bensch, R. Sher and J.D. Barchas, *Exp. Mol. Pathol.*, 1976, **25**, 38.
86 C. Heurteaux, J. Wissocq, T. Stelz and M. Thellier, *Biol. Cellulaire*, 1979, **35**, 251.
87 M. Thellier, T. Stelz and J. Wissocq, *J. Microsc. Biol. Cell*, 1976, **27**, 157.

88 J. Wissocq, T. Stelz, C. Heurteaux, J.C. Bisconte and M. Thellier, *J. Histochem. Cytochem.*, 1979, **27**, 1462.
89 M. Thellier, J. Wissocq and A. Monnier, *Lithium: Inorganic Pharmacology and Psychiatric Use*, ed. N.J. Birch, I.R.L. Press, Oxford, 1988, p. 271.
90 S.C. Nelson, M.M. Herman, K.G. Bensch and J.D. Barchas, *J. Pharmacol. Exp. Therap.*, 1980, **212**, 11.
91 B.S. Carpenter, D. Samuel, I. Wassermann and A. Yuwiler, *J. Radioanal. Chem.*, 1977, **37**, 523.
92 J. Wissocq, C. Heurteaux and M. Thellier, *Neuropharmacology*, 1983, **22**, 227.
93 C. Heurteaux, N. Baumann, F. Lachapelle, J. Wissocq and M. Thellier, *J. Neurochem.*, 1986, **46**, 1317.
94 R.J. Davie, *Lithium and the Cell*, ed. N.J. Birch, Academic Press, London, 1991, p. 243.
95 J.D. Phillips, R.J. Davie, M.R. Keighley and N.J. Birch. *J. Am. Coll. Nutr.*, 1991, **10**, 200.
96 C. Padgham, N.J. Birch, R.J. Davie and J.D. Phillips, *Magnesium and the Cell*, ed. N.J. Birch, Academic Press, London, 1993, p. 263.
97 R.J. Davie, J.D. Phillips, and N.J. Birch, *Trace 89 (Proceedings of the Third International Congress on Trace Metals in Health and Disease, Adana, Turkey*, 1989), eds. G.T. Yuregar, O. Donma and L. Kayrin, Cukerova University Publishing Company, Cukerova, Turkey, 1991, p. 325.
98 N.J. Birch, R.J. Davie and J.D. Phillips, *Magnesium – A Relevant Ion?*, eds. B. Lasserre and J. Durlach, John Libbey, London, 1991, p. 125.
99 N.J. Birch, *Handbook of Metal – Ligand Interactions in Biological Fluids*, ed. G. Berthon, Marcel Dekker, New York, 1994.
100 G.M.H. Thomas, M.S. Hughes, S. Partridge, R.I. Olufunwa, G. Marr and N.J. Birch, *Biochem. Soc. Trans.*, 1988, **16**, 208 (Abstract).
101 J.D. Phillips, M.S. Hughes and N.J. Birch, *Magnes. Res.*, 1990, **3**, 65 (Abstract).
102 S. Partridge, M.S. Hughes, G.M.H. Thomas and N.J. Birch, *Biochem. Soc. Trans.*, 1988, **16**, 205 (Abstract).
103 M.S. Hughes, *Lithium: Inorganic Pharmacology and Psychiatric Use*, ed. N.J. Birch, I.R.L. Press, Oxford, 1988, p. 285.
104 F.G. Riddell, *Lithium and the Cell: Pharmacology and Biochemistry*, ed. N.J. Birch, Academic Press, London, 1991, p. 85.
105 D. Mota de Freitas, J. Silberberg, M.T. Espanol, E. Dorus, A. Abraha, W. Dorus, E. Elenz and W. Whang, *Biol. Psychiat.*, 1990, **28**, 415.
106 F.G. Riddell, A. Patel and M.S. Hughes, *J. Inorg. Biochem.*, 1990, **39**, 187.
107 A. Abraha, E. Dorus and D. Mota de Freitas, *Lithium*, 1991, **2**, 118.
108 J. Bramham, *Lithium-7 NMR Investigations of the Biological Behaviour of the Lithium Ion*, PhD thesis, University of St Andrews, Scotland, 1993, p. 1.
109 F.G. Riddell and J. Bramham, *Lithium in Medicine and Biology*, eds. N.J. Birch, C. Padgham and M.S. Hughes, Marius Press, Carnforth, UK, 1993, p. 253.
110 M.E. Maguire, *Metal Ions in Biological Systems. Volume 26: Magnesium and Its Role in Biology, Nutrition and Physiology*, eds. H. Sigel and A. Sigel, Marcel Dekker, New York, 1990, p. 135.
111 F.W. Heaton, *Metal Ions in Biological Systems. Volume 26: Magnesium and Its Role in Biology, Nutrition and Physiology*, eds. H. Sigel and A. Sigel, Marcel Dekker, New York, 1990, p. 119.
112 J.J.R. Frausto da Silva and R.J.P. Williams, *Nature*, 1976, **263**, 237.
113 N.J. Birch, R.P. Hullin, R.A. Inie and F.C. Leaf, *Br. J. Pharmac.*, 1974, **52**, 132P (Abstract).

114 P.K. Kajda, N.J. Birch, M.J. O'Brien and R.P. Hullin, *J. Inorg. Biochem.*, 1979, **11**, 361.
115 P.K. Kajda and N.J. Birch, *J. Inorg. Biochem.*, 1981, **14**, 275.
116 M.S. Hughes, S. Partridge, G. Marr and N.J. Birch, *Magnes. Res.*, 1988, **1**, 35.
117 S.G. Brown, R.M. Hawk and R.A. Komoroski, *J. Inorg. Biochem.*, 1993, **49**, 1.
118 W.E.C. Wacker, *Magnesium and the Cell*, ed. N.J. Birch, Academic Press, London, 1993, p. 1.
119 W.R. Sherman, *Lithium and the Cell*, ed. N.J. Birch, Academic Press, London, 1991, p. 121.
120 R.H. Michell, *Biochem. Soc. Trans.*, 1989, **17**, 1.
121 M.J. Berridge, C.P. Downes and M.R. Hanley, *Biochem. J.*, 1982, **206**, 587.
122 M.J. Berridge, *JAMA*, 1989, **262**, 1834.
123 K. Nogimori, P.J. Hughes, M.C. Glennon, M.E. Hodgson, J.W.J. Putney and S.B. Shears, *J. Biol. Chem.*, 1991, **266**, 16499.
124 S. Cockcroft, B. Geny and G.M.H. Thomas, *Biochem. Soc. Tran.*, 1991, **19**, 299.
125 S. Cockcroft and G.M.H. Thomas, *Biochem. J.*, 1992, **288**, 1.
126 S.R. Nahorski, C.I. Ragan and R.A. Challiss, *Trends Pharmacol. Sci.*, 1991, **12**, 297.
127 A. Cornish-Bowden, *FEBS Lett.*, 1986, **203**, 3.
128 S.B. Shears, *Lithium: Inorganic Pharmacology and Psychiatric Use*, ed. N.J. Birch, I.R.L. Press, Oxford, 1988, p. 201.
129 M.J. Berridge, C.P. Downes and M.R. Hanley, *Cell*, 1989, **59**, 411.
130 N.J. Birch, *J. Trace Microprobe Tech,*. 1994, **12**, 1.
131 N.J. Birch, *Lithium and the Cell: Pharmacology and Biochemistry*, ed. N.J. Birch, Academic Press, London, 1991, p. 159.

CHAPTER 3

Gold complexes with Anti- arthritic, Anti-tumour and Anti-HIV Activity

C. FRANK SHAW III

Department of Chemistry, The University of Wisconsin-Milwaukee,
PO Box 413, Milwaukee, WI 53201, USA

1 Introduction

Chrysotherapy

In current medical practice, chrysotherapy, the treatment of rheumatoid arthritis with gold-based drugs, is well established. It derives its name from Chryses, a golden-haired heroine of Greek Mythology. Five gold(I) complexes (Figure 1) are widely used throughout the world in these treatments. Thiomalatogold(I), thioglucose gold(I), and thiopropanolsulfonate gold(I) are oligomeric complexes that contain linear gold(I) ions connected by bridging thiolate ligands. Bis(thiosulfate)gold(I) contains gold bound to the terminal sulfur donor atoms of $S_2O_3^{2-}$. The newest drug, auranofin, contains coordinated triethylphosphine and 2,3,4,6-tetra-O-acetyl-β-1-D-thioglucose ligands. The last two complexes are discrete monomers with well-defined structures; the oligomers, however, form non-crystalline powders and may have variable chain-lengths in solution. For AuSTm and AuSTg, the structures given in Figure 1 are the principal components of complex and somewhat variable mixtures that have been reviewed elsewhere.[1,2]

History of Medicinal Uses

Gold, the noblest of metals, was probably the first pure element recognized by the early inhabitants of the earth. It has a medicinal history that can be traced through the written history of every culture and far into pre-history by means of archeological records. Medicinal properties were early assigned to gold on the basis of its mystical importance and its association with the sun.[3] Modern medicine dismisses such uses for lack of plausible mechanisms. Yet, one can still

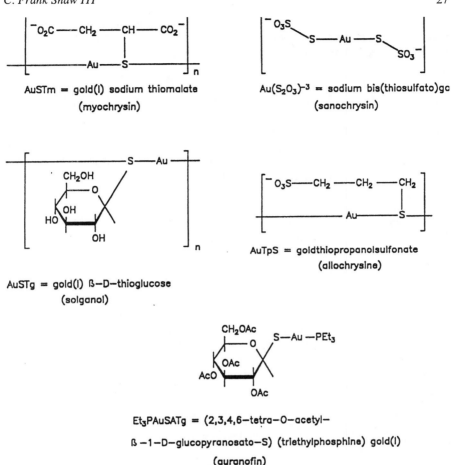

$AuSTm$ = gold(I) sodium thiomalate
(myochrysin)

$Au(S_2O_3)^{-3}$ = sodium bis(thiosulfato)go
(sanochrysin)

$AuSTg$ = gold(I) ß—D—thioglucose
(solganol)

$AuTpS$ = goldthiopropanolsulfonate
(allochrysine)

$Et_3PAuSATg$ = (2,3,4,6—tetra—O—acetyl—

ß —1—D—glucopyranosato—S) (triethylphosphine) gold(I)

(auranofin)

Figure 1 *Structures of five widely used gold (I)-based anti-arthritic drugs*

obtain modern descendants of such treatments. 'Goldwasser' and 'Goldschlager' are spiced liqueurs containing particles of elemental gold foil (about a dime's worth of gold in a litre) and reputed to have medicinal properties.

Chrysotherapy, the modern, scientific use of gold compounds in medicine, had its inception in the experimental work of Robert Koch, a German physician and bacteriologist. He discovered that $Au(CN)_2{}^-$ has potent bacteriostatic properties: that is, it retards bacteria from growing, but without killing them. This discovery led in the 1920s and 1930s to the use of gold compounds, primarily thiolate compounds (Figure 1), to treat tuberculosis. During the late 1920s, a Parisian physician, Jacques Forestier, noted that patients with arthritis who were taking gold had significant improvements, and he undertook trials of gold complexes as a treatment for rheumatoid arthritis. The efficacy of chrysotherapy was confirmed by large multinational trials just after World War II. Although chrysotherapy by rheumatologists has since undergone cycles of popularity and disuse, the nature of chrysotherapy treatment has changed very little except for the licensing of auranofin in the mid-1980s. Auranofin has the advantage of being orally absorbed, but now appears to be less effective than the traditional, injectable gold(I) thiolates.[4]

Rheumatoid Arthritis and Putative Mechanisms of Action

Rheumatoid arthritis, for which chrysotherapy is used world-wide, is an inflammatory condition that leads to progressive erosion of the articular cartilage lining the interfaces of bones in joints.[5] If the attack, which shows many characteristics of an autoimmune disease, is not checked, the bones will eventually fuse after complete loss of the cartilage. The initial inflammation occurs in the synovial membrane, which surrounds the joints, and then moves into the synovial cavity between the bones. Damage to the tissues is effected by lysosomal enzymes including collagenase and other proteases that are released because of the inflammatory condition. The resulting tissue destruction releases cell and tissue fragments, which stimulate further inflammation and the migration of immune cells including macrophages into the inflamed area. Thus, a cycle of degradation and further release of destructive enzymes is set up.

Chrysotherapy is effective for about 70% of the patients taking the treatment.[1,5] Others do not benefit from gold, or suffer side effects that require cessation of treatment. Most side effects are mild, and some are only cosmetic (*e.g.* skin rashes). Only occasionally are there life-threatening consequences (*ca.* 1:10,000 patients): usually inhibition of white or red blood cell formation in the bone marrow. Patients are usually monitored weekly or monthly for blood gold levels, which are typically held below $300 \mu g \, dl^{-1}$ ($\sim 15 \mu M$) to minimize the accumulation of gold in tissues and the resulting side effects. The injectable gold(I) thiolates (AuSTg, AuSTm, *etc.*) have activity superior to that of auranofin ($Et_3PAuSATg$).[5]

Gold drugs have well-documented anti-inflammatory activity, which they share with many organic medicines such as aspirin and ibuprofen, but they also have more profound effects on the underlying joint destruction that simple anti-inflammatory agents lack. Thus some patients on chrysotherapy obtain long-lasting remission of the disease, which is sustained with maintenance therapy, less frequent treatment with the gold drug. Many mechanisms have been proposed for chrysotherapy, but none can be considered as well-established within the medical community. The proposed mechanisms listed in Table 1 range from broad and sometimes vague systemic effects, such as immunomodulatory activity, to very specific biochemical processes that act through a progressively multiplied effect on a larger system. One problem in examining the mechanisms of chrysotherapy has been a general failure to recognize that gold complexes are in fact pro-drugs. They undergo rapid metabolism to form protein and low molecular-weight metabolites that are probably the source of gold for the elusive target sites. As discussed elsewhere in this review, the gold bound to albumin is less bio-available than are the drugs themselves, and less uptake into cells occurs in its presence. This situation is likely to prevail

Table 1 *Possible mechanisms of chrysotherapy*

Anti-inflammatory activity
  Altered prostaglandin biosynthesis
Inhibition of lysosomal enzymes activity and/or release
Immunomodulatory activity
  Altered peptide presentation
  Inhibited macrophage phagocytosis
  Inhibited histamine release by mast cells
  Inhibited complement activity
  Inhibited monocyte migration
  Inhibition T-cell stimulation and proliferation
  Reduced immunoglobin (including rheomatoid factor) titres
Altered copper homeostasis
Altered metabolism of reactive oxygen species
  Catalysis of singlet-to-triplet oxygen conversion
  Glutathione peroxidase inhibition
  Modulation of the oxidative burst
Perturbation of Zn-fingers by Au(I) thiol competition

---

in vivo. On the other hand, $Au(CN)_2^-$, which has been identified as a common metabolite of all gold drugs, is more readily taken into cells. It occurs at concentrations of 1–5 ppb ($\sim$ 5–25 nM) in blood and at higher concentrations in urine.[6]

Some drugs bind to their target sites so tightly that radiolabelling leads to quick identification of the molecular mechanism of action. Gold unfortunately is rather promiscuous and associates with many tissues, cells, and proteins in the body. The concentrations are higher in inflamed joints than in the blood, but only marginally so: 2–3-fold. Only when a mechanism of action is well established will it be possible to design new (third-generation) drugs that combine the effectiveness of the (first generation) injectable gold(I) thiolates with the safety of auranofin (the only second generation drug).

2 Gold Chemistry

Oxidation States

Gold can exist in seven oxidation states: − I, 0, I, II, III, IV, and V. Apart from gold(0) in the colloidal and elemental forms, only gold(I) and gold(III) are known to form compounds that are stable in aqueous media, and hence in a biological milieu. Relativistic effects arising from the interaction of the large nuclear charge (+ 79) affect the 6s orbital wavefunction and partially determine the properties of gold atoms, ions and complexes. Major review articles by Sadler,[7] Shaw,[8] and Brown and Smith[9] detail the subtleties of gold chemistry in relationship to biological systems. Puddephatt[10] surveys the basic features of gold chemistry. Extensive surveys of the structural chemistry of gold compounds were published in 1983 by P. G. Jones[11] and in 1986 by Melnick and Parish.[12] Only the broad outlines of gold(I) and gold(III) chemistry will be reviewed here.

Gold(I) Complexes

Gold(I) is a d^{10} metal ion. Like Ag(I), Cu(I), and Hg(II), it forms complexes that may be two-, three-, or four-coordinate, with various combinations of ligands. Typical examples are shown in Figure 2. Gold(I) has much less tendency than Cu(I) and Ag(I) have to adapt 3- or 4-coordination. Thus $Cu(CN)_n^{-n+1}$ and $Ag(CN)_n^{-n+1}$ are known for $n = 2$, 3 and 4; for gold, however, only $Au(CN)_2^-$ has been isolated or characterized, despite many efforts to form the corresponding tri-and tetra-cyano complexes.

Two-coordinate complexes may be generally be obtained with any type of ligand able to coordinate to soft metal centers. These complexes may be neutral, positive or negative, and are linear, with bond angles approaching 180°. Thiolates (such as cysteine and thiomalate), thioethers (such as methionine and dimethylsulfide), selenols, phosphines, cyanide and alkyl groups (especially

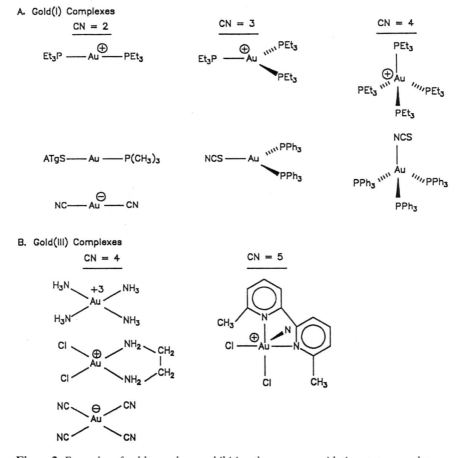

Figure 2 *Examples of gold complexes exhibiting the common oxidation states, I and III*

methyl and phenyl), and a host of nitrogen donor ligands (including amines and heterocycles) form strong bonds and stable complexes. Oxygen complexes are known, but are generally reactive and much less stable (*e.g.* $[(Ph_3PAu)_3O^+]$, used by Schmidbaur and co-workers to prepare hypervalent gold cluster complexes[13]). Three-coordinate complexes are trigonal planar and generally contain at least one and often two neutral ligands such as phosphines or arsines.[14] No trithiolate complexes of gold have been isolated to date, even though these ligands have very high affinities for gold(I). Four-coordinate gold(I) complexes are also typically rich in phosphines and usually neutral or cationic.[14] The three- and four-coordinate d^{10} complexes preferentially adapt trigonal planar and tetrahedral arrangements, respectively, of the ligands about gold.

Gold(III) Complexes

Gold(III) is a d^8 metal ion, a configuration it shares with Pt(II) and Rh(I). Thus gold(III) complexes, with a few notable exceptions, are generally four-coordinate and square planar (Figure 2). A variety of ligands form stable complexes with this oxidation state; as a result, its complexes have a wide range of physical and chemical properties. With four neutral ligands, such as NH_3, the charge is $+ 3$; with four anions such as Cl^-, the charge is $- 1$; from appropriate combinations any intermediate charge can be obtained. The ability to modulate charge is important for the design of complexes with the appropriate balance of hydrophilicity and lipophilicity for medicinal use or testing.

Oxidation–Reduction Potentials

Neither gold(I) nor gold(III) forms a stable aquated ion ($[Au(OH_2)_{2-4}^+]$ or $[Au(OH_2)_4^{3+}]$, respectively) analogous to those found for many transition metal and main group cations. Both are thermodynamically unstable with respect to elemental gold and can be readily reduced:

$$Au^{1+} + 1e^- = Au^0 \qquad E_0 = + 1.68 \qquad (1)$$

$$Au^{3+} + 3e^- = Au^0 \qquad E_0 = + 1.42 \qquad (2)$$

The large, positive E_0 values indicate that mild reducing agents are able to effect the reduction. For example, even water slowly reduces some gold complexes (*e.g.* $AuCl_4^-$), with concomitant release of elemental oxygen. The situation can easily be changed by appropriate choice of ligands to stabilize the oxidized forms of gold; for example:

$$Au(SCy)_2^- + 1e^- = Au^0 + 2CyS^- \qquad E_{0,\frac{1}{2}} = - 0.14 \qquad (3)$$

$$AuBr_4^- + 3e^- = Au^0 + 4Br^- \qquad E_0 = + 0.858 \qquad (4)$$

The E_0 value for $AuBr_4^-$ is significantly less than that for free Au^{3+}, but indicates that $AuBr_4^-$ still is a powerful oxidant. For the $Au(SCy)_2^-$ complex (CyS = cysteinato), the negative potential demonstrates significant stabilization of the $+1$ oxidation state by coordination to thiolate ligands. Thus, gold(III) tetrahalide complexes remain powerful oxidizing agents, but gold(I) can be stabilized by cyanide and thiolate ligands.

Reduction of gold(III) to gold(I) or gold(0) is often observed in biological milieux. The reaction can be driven by naturally occurring reductants such as thioethers, thiols or even disulfides:[15-17]

$$Au^{III}L_4 + RSR + H_2O \rightarrow Au^IL_2 + RS(O)R + 2L + 2H^+ \qquad (5)$$

$$Au^{III}L_4 + 2RSH \rightarrow Au^IL_2 + RSSR + 2L + 2H^+ \qquad (6)$$

$$mAu^{III}L_4 + \tfrac{1}{2}RSSR + nH_2O \xrightarrow{m=(2n-1)/3} mAu^I + RSO_n^- + 4mL + 2nH^+ \qquad (7)$$

The exchange of monodentate ligands bound to gold(I) or gold(III) is generally rapid and proceeds by associative mechanisms that involve, respectively, three-coordinate gold(I) and five-coordinate gold(III) transition states (and occasionally intermediates). The presence of bidentate, tridentate or macrocyclic ligands reduces the thermodynamic tendency for ligand exchange and often leads to slower reactions even when they are favorable.

3   Gold Biochemistry and Pharmacology

The gold-based anti-arthritic agents are really pro-drugs that undergo rapid metabolism to form new metabolites. During the early 1990s, a number of exciting observations provided new insights into the chemical transformations that transpire *in vivo*.[18] These are discussed in several of the following sections, along with summaries of previous research that provides a basis for understanding the latest findings. Following the early triad of in-depth reviews on the bio-inorganic chemistry of gold,[7-9] general reviews by Parish and Cotrill,[19] Smith and Reglinski[20] and Grootveld *et al.*,[21] have appeared, in addition to specialized reviews on cellular pharmacology by Crooke *et al.*,[22] EXAFS, XANES and WAXS structural studies by Elder and Eidsness[23] and studies on protein chemistry by Shaw[24] and on aurothioneins (gold-laden metallo-thioneins) by Savas and Shaw.[25]

In Vivo Metabolism and Ligand Displacement

Biological studies in man and in laboratory animals show that the gold drugs used clinically undergo rapid metabolism *in vivo*. The use of radiolabelled ligands reveals that gold and its carrier ligands have different distributions and excretion times.[26-28] The phenomenological term 'dissociation' that is sometimes applied to this finding is inapt, because cellular and biochemical studies demonstrate that ligand exchange reactions are the primary mechanism of metabolism and that free gold ions do not result. For that reason, ligand displacement is the preferred terminology. Rapid displacement of the ligands from $^{198}Au^{35}STg$ in mice and $^{198}Au^{35}STm$ in rats has been reported.[27,29] The retention of ^{198}Au in various organs is much greater than the retention of ^{35}S label from the thiolate ligands. Likewise, the three components of radiolabelled auranofin $(Et_3^{32}P-^{195}Au-^{35}SATg)$ are metabolized differently in dogs (Table 2): the half lives for excretion of $Et_3^{32}P=O$ and ^{35}S are 8 and 16 hours, respectively, while the half-life for gold excretion is 20 days.[28] The relevance of the results to humans is clear from clinical results that detected free thiomalate (TmSH) in the blood of patients receiving myochrysine (AuSTm).[26]

Studies of auranofin added to whole blood (Table 2) confirmed that the three components of the drug have very different fates and showed that the ligand displacements are dramatically faster than the excretion rates quoted above.[28] Within 20 min the gold is primarily protein-bound in the serum, and the triethyl-phosphine is distributed among the red cells (2:5), proteins (1:5), and a low-molecular-weight species (2:5) known to be phosphine oxide (Et_3PO).[28]

Cellular Chemistry – The Thiol Shuttle Model

The injectable gold drugs such as AuSTm and AuSTg are not readily taken up by most cells. Yet it is clear that they can bind to cell surface thiols and by such a mechanism may affect the cells' overall metabolism.[30,31] As discussed in a subsequent section, cyanide alters the situation and facilitates gold uptake in red cells.

Auranofin, in contrast to the polymeric thiolates, is extensively and quickly taken up by various cell types.[28,32] The data for red cells in Table 2 show that the acetylthioglucose ligand is displaced before gold is transported into cells, while the triethylphosphine enters the cell but does not accumulate to the same extent as the gold itself.[28] Mirabelli and co-workers developed a sulfhydryl shuttle model for the uptake and efflux of Et_3PAu^+ from auranofin, based on studies in cultured macrophage cells.[32] Figure 3 shows the principal features of the model, in which sulfhydryl-dependent membrane transport proteins (MSH) provide a vehicle for movement of the Et_3PAu^+ across the cell membrane. Within the cell, Et_3PAu^+ is transferred to other cell sulfhydryls (CSH) and can then undergo several fates. Further reaction with additional cell sulfhydryls

Table 2 *Ligand displacement from radio-labelled auranofin: $Et_3^{32}P-^{195}Au-^{35}SATg$*

| Component | Rat blood in vitro[a,b] | | | Dog blood/urine in vivo[a,c] | | |
	% Red cells	% Protein	% LMW	t_{max}	$t_{1/2}$	% Excreted[d]
^{198}Au	65 ± 3	35 ± 3	$\ll 1$	8 h	20 d	6
^{35}S-AtgSH	0	82 ± 1	19 ± 1	2 h	16 h	81
^{32}P-Et_3P	38 ± 8	19 ± 1	43 ± 7	1.5 h	8 h	90

[a]Ref. 28. [b]20 min incubation (LMW = low molecular weight fractions). [c]0–200 $\mu g \, kg^{-1}$ oral dose. [d]72 hour total excretion.

SULFHYDRYL SHUTTLE MODEL

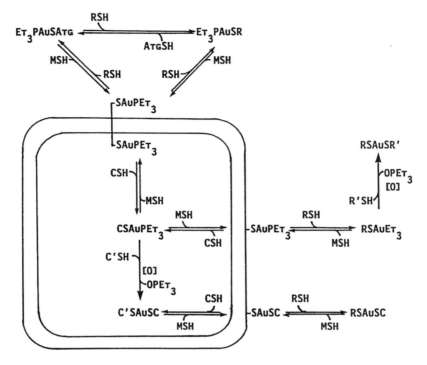

Figure 3 *The Sulfhydryl Shuttle Model (adapted from Ref. 32). MSH = membrane transport-protein thiols; CSH, C'SH = cellular thiols (protein and non-protein); RSH = extracellular thiols*

(C'SH) may lead to displacement of the Et_3P ligand and its oxidation to $Et_3P=O$. With or without the Et_3P bound, the gold(I) can be shuttled out of the cell *via* the membrane sulfhydryl proteins.[32,33] Although the uptake is mediated by the membrane sulfhydryls, it is not energy-dependent, active transport. Therefore, intracellular gold concentrations should be in equilibrium with the extracellular sources of gold. This was tested in two ways. Serum albumin, which has a relatively high affinity for gold, was added to the media of cultured macrophages[34] and B16 melanoma cells[35] in increasing amounts. This led to slower and less extensive uptake of the gold, consistent with an equilibrium distribution. The cytotoxicity to tumour cells of both thiotheopyllinato (triphenylphosphine)gold(I) and auranofin is reduced by fetal calf serum, which contains serum albumin, which reduces the uptake.[35,36] An equilibrium must be reversible, and red cells that had accumulated gold upon exposure to auranofin were shown to have more extensive gold efflux in the presence of added serum or just albumin than in isotonic salt solution.[33] These findings strongly support the existence of an equilibrium between intra- and extra-cellular auranofin metabolites.

Protein Chemistry

The high affinity of gold(I) for sulfur[7–9,21,23,24] and selenium ligands[37] suggests that proteins, including enzymes and transport proteins, will be critical *in vivo* targets. In addition, it is clear that extracellular gold in the blood is primarily protein bound, suggesting protein-mediated transport of gold during therapy. The sections below summarize the bio-inorganic chemistry of several well-studied protein and enzyme systems that interact with gold complexes.

Serum Albumin

Serum albumin (Figure 4) is the major extracellular protein in the blood. It is held together by seventeen disulfide bonds and has an additional unpaired cysteine residue. Cys-34 is in the reduced form in mercaptalbumin, AlbSH, which comprises about 60–70% of the circulating albumin molecules, and is paired with cysteine or glutathione (AlbSSCy or AlbSSG) in the remainder. Albumin consists of three structurally related domains (I, II, & III), each with two

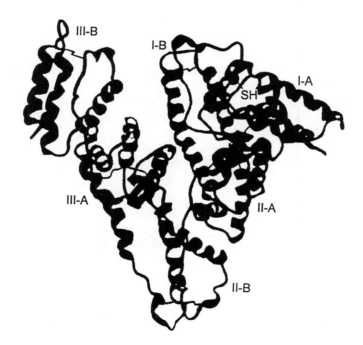

Figure 4 *The structure of human albumin, showing the domains, subdomains and cys-34. Cystein 34 is below the surface at the bend between helices h2 and h3 of domain IA. This is consistent with earlier literature establishing that it is protected from the solvent and only accessible due to motions previously described as crevice opening*
(Adapted from X. M. He and D. C. Carter, *Nature*, 1992, **358**, 209)

subdomains (A and B).[38] Cys-34 is located in domain I-A and lies buried below the surface of the protein.

Extensive bio-inorganic research has established that the principle gold(I) binding site is cys-34. AuSTm forms a complex in which gold is coordinated by cys-34 and thiomalate, with an average Au–S distance of 228 $\pm$ 2 pm:[39]

$$AlbS^- + 1/n[AuSTm]_n \rightarrow AlbS\text{–}AuSTm \qquad (8)$$

Similar complexes where glutathione, acetylthioglucose or thioglucose replaces the thiomalate have been isolated and characterized.[39-40] Auranofin reacts at cys-34 *via* a ligand exchange reaction that displaces the sulfhydryl group:[40-43]

$$AlbS^- + Et_3PAuSATg \rightarrow AlbS\text{–}Au\text{–}PEt_3 + ATgSH \qquad (9)$$

The Au–P and Au–S bond lengths of the phosphine-gold-albumin complex are 229 and 227 pm, respectively.[40] Under conditions approximating those *in vivo*, complex formation is first-order in auranofin and has a rate constant of $2.9 \pm 0.2\,s^{-1}$, which indicates that auranofin (and probably its deacetylated metabolite) will have a very short lifetime after entering the bloodstream where albumin is present in large excess ($\sim$ 400 μM AlbSH and $\sim$ 10 μM Au).[44] Sadler and colleagues have recently discovered a conformational change in albumin that accompanies gold binding to cys-34 (but not disulfide formation at cys-34).[45] The rate of gold binding may correspond either to the rate of opening of the cys-34 crevice to solvent molecules or to the rate of this recently found structural change that accommodates gold binding.

The free acetylthioglucose liberated *via* equation 9 reacts further with the cysteine 34 disulfide bonds to liberate cysteine[46] and also displaces the Et$_3$P ligand, leading to its oxidation:[47]

$$AlbS\text{–}SCy + ATgSH \rightarrow AlbSH + ATgS\text{–}SCy \qquad (10)$$

$$AlbS\text{–}Au\text{–}PEt_3 + ATgSH \rightarrow AlbS\text{–}Au\text{–}SATg + PEt_3 \xrightarrow{\;[O]\;} Et_3P{=}O \qquad (11)$$

Physiological thiol ligands such as glutathione (GSH) also drive the displacement of phosphine (equation 12).[47] The oxidant in equation 11 can be either molecular oxygen (equation 13a) or the albumin disulfide (equation 13b), yet the reaction proceeds at a similar rate and to about the same extent aerobically or anaerobically.[48] Since free O$_2$ is not present in serum, it is likely that disulfide bonds are the *in vivo* oxidants.[48]

$$AlbS\text{–}Au\text{–}PEt_3 + GSH \rightleftharpoons AlbS\text{–}Au\text{–}SG + PEt_3 \qquad (12)$$

$$PEt_3 \begin{cases} + 1/2O_2 \rightarrow Et_3P{=}O & (13a) \\[2em] + \underset{S}{\overset{S}{|}}{>}Alb\text{–}X + H_2O \rightarrow (HS)_2\text{–}Alb\text{–}X + Et_3PO & (13b) \end{cases}$$

Structure function studies that substituted Me_3P, Pr^i_3P or $(NCEt)_3P$ for Et_3P of auranofin in these reactions have established that the basicity of the phosphine controls the rate and extent of R_3PO formation:[48-50]

$$Me_3P \approx (NCEt)_3P > Et_3P \gg iPr_3P$$

Structure–function studies of the anionic ligand (Figure 5) revealed, surprisingly, that replacement of the AtgSH ligand by CN^- or the selenium analogue, AtSeH, which bind more tightly to gold than the thiol does, did not inhibit the first reaction (equation 9) and unexpectedly led to more rapid formation of Et_3PO (equation 11) than is observed with auranofin itself.[48,49]

Hemoglobin

In red cells exposed to 7 mM Et_3PAuCl, the Et_3PAu^+ moiety crosses the membrane and reacts with hemoglobin and glutathione.[43] Hemoglobin (Hb) is

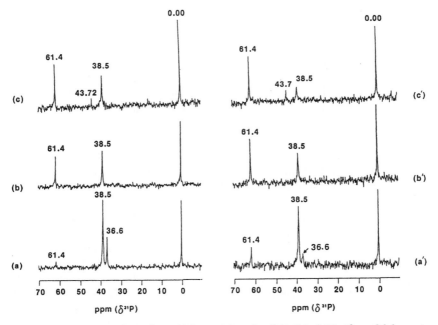

Figure 5 *Formation of Et_3PO ($\delta_P = 61.4$ ppm) from (a,a') $Et_3PAuSATg$ ($\delta_P = 36.6$ ppm) (b,b') Et_3PAuCN ($\delta_P = 35.4$ ppm) and (c,c') $Et_3PAuSeATg$ ($\delta_P = 37.7$ ppm) during reaction with serum albumin under argon; $AlbSAuPEt_3$ (38.5 ppm) and $(Et_3P)_2Au^+$] (43.8 ppm) (a,b,c) 1.5 hours and (a',b',c') 24 hours after mixing 4.05 mM BSA and 2.6 mM gold complex. For auranofin (a,a'), free drug and protein complex are present with only a trace of Et_3PO. For the cyanide (b,b') and tetraacetyl-β-1-D-selenoglucose complexes, the formation of Et_3PO is much greater at both time points. This demonstrates that increasing the affinity of the anionic ligand for gold(I) will accelerate rather than retard the rate of metabolism of phosphinegold(I)-based drugs*

an $\alpha_2\beta_1$ tetramer, which contains six cysteine residues (cys-α-104; cys-β-93; and cys-β-112). Only the cys-β-93 residues, which are near the surface, bind to Et_3PAu^+. The resulting complexes, $Hb-(SAuPEt_3)_n$, where $n \leq 2$, are not as robust as the $AlbSAuPEt_3$ complex described above. The hemoglobin reacts completely with Et_3PAuCl, to a smaller extent with $Et_3PAuSTg$, and negligibly with $Et_3PAuSATg$, indicating that the Hb cys-β-93 residues are lower affinity thiols than are AtgSH and AlbSH.[53] Consistent with this finding, there is efficient interprotein transfer of Et_3PAu^+ from hemoglobin to albumin:[53]

$$Hb-(SAuPEt_3)_{1.7} + xsAlbSH \rightarrow AlbSAuPEt_3 + Hb(SH)_2 \qquad (14)$$

It is interesting to note that this transfer occurs spontaneously and that low-molecular-weight thiols are not required to mediate the process.

Aurothioneins

Metallothionein (MT) is a ubiquitous metal-binding protein found in all verte-brate and invertebrate organisms.[54] MTs in mammals and crustaceans bind seven and six metal ions, respectively. Mammalian MTs utilize 20 cysteine residues among about 61 amino acids to group their seven Zn^{2+} or Cd^{2+} ions into $M_4(SCy)_{11}$ and $M_3(SCy)_9$ clusters, which are in domains designated α and β and are formed at the *C*- and *N*-terminal ends of the protein, respectively. Figure 6 shows the two clusters. Biological functions of MT include the detoxification of the environmentally accumulated Cd^{2+} ions, the mediation of Zn^{2+} and Cu^+ metabolism and, putatively, the detoxification of reactive oxygen species. Inter-estingly, adventitious reactions with certain medicinal agents, including the anti-tumour drugs chlorambucil (a mustard derivative) and cisplatin[55,56] and the gold-based anti-arthritic agents,[57,58] mediates their cytotoxicity. A variety of animal experiments summarized elsewhere[25] demonstrate that in the kidneys and liver, gold is bound to metallothioneins, along with Zn^{2+} and Cu^+, after administration of gold drugs. The bio-inorganic chemistry of gold incorporation into MT is described below.

Gold thiomalate displaces Zn^{2+} and Cd^{2+} from the protein in a biphasic reaction. Zn^{2+} is displaced in preference to Cd^{2+} when gold is limiting, and complete displacement of both metal ions is possible when excess AuSTm is present.[59] The kinetics are biphasic, with rate constants of $k_f = 2.4 \pm 0.5 \times 10^{-2}$ and $k_s = 9.6 \pm 1.7 \times 10^{-3}$. These reaction rates are much greater than are the rates of gold accumulation in the kidney or liver, and therefore demonstrate that the reaction with MT is not rate-limiting for accumulation.[59] Recent studies with isolated α- and β-domains show that the fast step is associated with the α-domain and the slow step with the β-domain.[60,61]

The resulting aurothioneins have been characterized by a variety of chromato-graphic and spectroscopic techniques. When gold is limiting, it loses the thiomalate ligand and binds to two cysteine sulfhydryls with $d_{Aus} = 229$ pm:[59]

$$Au^{35}STm + xsZn,Cd-MT \rightarrow Au,Zn,Cd-MT + Tm^{35}SH + Zn^{2+} \qquad (15)$$

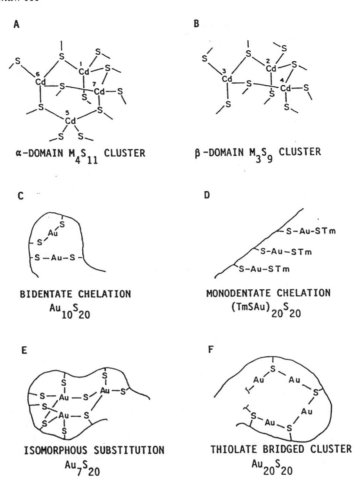

A

α-DOMAIN M_4S_{11} CLUSTER

B

β-DOMAIN M_3S_9 CLUSTER

C

BIDENTATE CHELATION
$Au_{10}S_{20}$

D

MONODENTATE CHELATION
$(TmSAu)_{20}S_{20}$

E

ISOMORPHOUS SUBSTITUTION
Au_7S_{20}

F

THIOLATE BRIDGED CLUSTER
$Au_{20}S_{20}$

Figure 6 *Metallothionein metal coordination. (A,B) The four- and three-metal clusters of*
M_7-MT where M = Zn^{2+} or Cd^{2+} with tetrahedral MS_4 coordination. (C,D)
Bidentate and monodentate chelation of Au(I) by metallothionein as observed in
Cd,Zn,Au-MT and (TmSAu)20 MT, respectively. (E,F) Isomorphous substitution
and thiolate-bridged coordination of gold(I) in MT, which have not been reported
to date. The formulae under structure C–F are the limiting cases for saturation of
the protein by each mode of gold coordination

When gold is present in excess, the thiomalate ligand is retained and the initially
bound Cd^{2+} and Zn^{2+} are completely displaced:

$$xsAu^{35}STm + Zn,Cd\text{-}MT \rightarrow (Tm^{35}SAu)_{\sim 20}MT + 7M^{2+} \qquad (16)$$

In the latter case, the stoichiometry of bound gold approaches the number of
available sulfhydryls, 20 per MT, and the bond-lengths are similar,
$d_{AuS} = 230$ pm. In both reaction products, the gold remains gold(I); no evidence

for elemental Au or Au(III) is obtained. Four possible coordination models are shown in Figure 6. From the AuS_2 coordination environment, the stoichiometries of metal exchange and the retention or loss of the thiomalate ligand, it is clear that Au, Zn, Cd-MT contains Au(I) in bidentate coordination and $(Tm^{35}SAu)_{\sim 20}MT$ in monodentate coordination.[59] In either case, the two-coordination requirement of the gold(I) must distort the cluster structure, and it completely unfolds the tertiary structure in $(TmSAu)_{\sim 20}MT$.

Auranofin does not react *in vitro* with MT,[41,62] but gold does bind to MT *in vivo* and in cultured cells after gold drugs have been administered. The apparent discrepancy reflects the rapid *in vivo* metabolism of auranofin, leading to altered ligation of the gold. Et_3PAuCl, in which the less tightly bound Cl^- ion is present, reacts more readily and displaces bound Zn^{2+} from Cd, Zn-MT. MT induced by auranofin is degraded more rapidly than it is after Zn^{2+} or Cd^{2+} induction (Table 3), an effect that is attributed to the altered protein structure that results from two-coordination of Au(I) to the sulfhydryl groups.[63] The potential interaction of gold complexes with MT *in vivo* may reduce their cytotoxicity, necessitating larger doses than would be needed in the absence of this protein.

Glutathione Peroxidase

Elegant biochemical and animal studies by Tappel and co-workers[64,65] have described the interaction of gold thioglucose with the selenium-dependent enzyme, glutathione peroxidase (GPx-ase). This enzyme is essential to red cell integrity, because it detoxifies H_2O_2 formed fortuitously. In addition, H_2O_2 is formed by the oxidative burst at inflamed sites, where surrounding cells can also be protected by the peroxidase activity. Two equivalents of glutathione (GSH, a thiol which has important functions as a biological reductant) reduce H_2O_2 into water and, in turn, are oxidized to the corresponding disulfide, GSSG (equation 17a).

$$2GSH + H_2O_2 \rightarrow GSSG + 2H_2O \tag{17a}$$

The active site of GPx-ase contains a selenol group (SeH) which is oxidized to Se(=O)H with formation of one equivalent of water in the first step of the enzymatic cycle (equation 17b). Two equivalents of glutathione then react with

Table 3 *Biological half-lives of MT with various metal loadings*[a]

Metal ion	$T_{1/2}$ (h)
Cd^{2+}	24
Zn^{2+}	10
$Au(I)^b$	0.75

[a]Measured in cultured Chinese hampster ovary cells after MT induction with the indicated metal ion. Ref. 63. [b]MT induction with auranofin.

the oxidized selenium. The first forms an Se–S bond (Gpx–Se–SG) with release of the second water molecule. Finally, the second GSH attacks the sulfur to release the glutathione disulfide and restore the selenol.

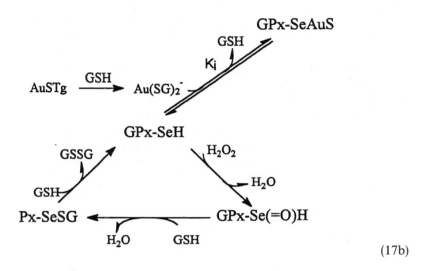

(17b)

The active site selenol has a high affinity for gold binding and is the site of inhibitory action (equation 17b). Tappel determined that the inhibition constant (K_i) for gold(I) thioglucose (AuSTg) has a value of 2.3 μM, which can be interpreted as the concentration of gold which will reduce the enzyme activity by a factor of one-half. The enzyme assay uses 1 mM GSH for the reaction and Tappel et al.[64,65] suggested that the gold(I) thioglucose was converted in situ to bis(glutathionato)gold(I), and this has now been confirmed by Roberts[66] who showed that authentic Au(SG)$_2{}^-$ exhibits the same K_i value, 2.3 μM. For the same reason, Et$_3$PAuCl and Et$_3$PAuSATg have similar K_i values, 10 ± 1 μM, because they are converted into Et$_3$PAuSG by in situ reaction with GSH.[66] The Et$_3$P ligand has a stronger affinity than GSH for gold(I) and is not displaced to form Au(SG)$_2{}^-$ in the assay mixture. These reactions show that care is required in interpreting the chemistry of gold(I) compounds in biological milieux, because they undergo rapid ligand exchange reactions with thiols and other high affinity ligands that may be present.

The K_i values of 2.3 and 10 μM for gold(I) thiolates and triethylphosphinegold(I) complexes, respectively, are below the 5–15 μM concentrations of gold observed in patients undergoing chrysotherapy and, therefore inhibition in vivo is likely. In contrast, many other enzymes which are inhibited by gold(I) have K_i values in the millimolar range or higher and are unlikely to be inhibited in vivo. Tappel[65] has suggested that inhibition of GPx-ase could lead to a buildup of H$_2$O$_2$ at inflammatory sites, leading to cell toxicity that would slow the immune response and contribute to clinical improvement for the patients.

Cyanide Metabolites

Evidence accumulating over the past decade has established the importance of cyanide complexes, principally $Au(CN)_2^-$, in the metabolism of gold drugs. $Au(CN)_2^-$ also has anti-HIV activity, as discussed later in this chapter. The first evidence of $Au(CN)_2^-$ as a metabolite was the report that tobacco smoking enhances the ability of patients' red cells to accumulate gold metabolites of AuSTm and AuSTg, although animal red cells do not.[67,68] The inhaled smoke of tobacco products contains up to 1700 ppm of HCN, which is absorbed through the lungs. It can react with gold and will facilitate the transfer of gold into red cells and presumably other cells.[67] [For obvious reasons this phenomenon was not observed in laboratory animals, whose life spans are briefer than the minimum age for cigarette purchases.] The involvement of cyanide also explains the higher incidence of side effects from gold drugs in patients who smoke.[68,69]

Several studies have examined the chemistry of thiol–cyanide exchanges, which follow the general equations below:[70–72]

$$1/n\,[AuSR]_n + HCN \rightleftharpoons RSAuCN^- + H^+ \qquad (18)$$

$$Au(SR)_2^- + HCN \rightleftharpoons RSAuCN^- + RSH \qquad (19)$$

$$2\,RSAuCN^- \rightleftharpoons Au(CN)_2^- + Au(SR)_2^- \qquad (20)$$

$$RSAuCN^- + HCN \rightleftharpoons Au(CN)_2^- + RSH \qquad (21)$$

The equilibria depend on pH, the affinity of the thiol(ate)s for gold(I), and the ratios of thiol and cyanide to gold that are present in the system under study. Elder has measured two apparent equilibrium constants:[73]

$$1/n\,[AuSTm]_n + 2\,HCN \xrightarrow{\;K_{app}7.4\,=\,6 \times 10^2\,M^{-1}\;} Au(CN)_2^- + TmSH \quad (22)$$

$$Au(CN)_2^- + 2\,CySH \xrightarrow{\;K_{app}7.4\,=\,6 \times 10^{-3}\;} Au(SCy)_2^- + 2\,HCN \quad (23)$$

Traces of $Au(CN)_2^-$ have been identified as a common metabolite of all three gold drugs (AuSTm, AuSTg, and auranofin) used clinically in the United States.[6,73] Its presence is independent of the smoking habits of the patients, but the concentration is somewhat higher in smokers. Research on the effect of gold compounds on the immune system led to the finding that thiocyanate (SCN^-) can be converted into cyanide by hypochlorite generated during the oxidative burst of immune cells.[69]

$$SCN^- \xrightarrow{\;OCl^-\;} CN^- \xrightarrow{\;Au(SR)_2^-\;} Au(CN)_2^- \qquad (24)$$

This suggests that formation of $Au(CN)_2^-$ may be a critical metabolite at the inflammatory sites in the arthritic joints of patients. The $Au(CN)_2^-$ is taken up into the cells and can limit the extent of the oxidative burst.[69] The mechanism of $Au(CN)_2^-$ transport into red cells appears to be via the sulfhydryl shuttle mechanism (see Figure 3).[63] Intact $Au(CN)_2^-$ ions bind to serum albumin at one strong and three weak binding sites, although only the former is physiologically significant.[74,75] The stronger interaction is surprisingly large, but the

$$Alb + Au(CN)_2^- \xrightleftharpoons[n = 0.8]{K_S = 5.5 \times 10^4 \text{ M}^{-1}} Alb \cdot [Au(CN)_2^-]_S \qquad (25)$$

$$Alb \cdot Au(CN)_2^- + Au(CN)_2^- \xrightleftharpoons[n = 3]{K_W = 7 \times 10^3 \text{ M}^{-1}} Alb \cdot [Au(CN)_2^-]_S \cdot [Au(CN)_2^-]_W \qquad (26)$$

labile binding is easily reversed and thus is consistent with cellular uptake of $Au(CN)_2^-$ in the bloodstream.[70]

These newly found roles of $Au(CN)_2^-$ may open the door to new research and to studies on the mechanisms of action that will end the long uncertainty about where and how the gold drug metabolites exert their effects on the inflamed joints and/or immune systems of patients.

Oxidation States *In Vivo*

Chemical reactions of gold drugs exposed to body fluids and proteins are predominantly ligand exchange reactions that preserve the gold(I) oxidation state,[8,19,33,76] exemplified by the protein reactions described above. A considerable body of evidence suggests that *in vivo*, gold exists, and would be expected to exist, primarily as gold(I). Aurosomes (lysosomes that accumulate large amounts of gold and undergo morphological changes) taken from gold-treated rats contain predominantly gold(I), even when gold(III) has been administered.[23,76] Methionine in proteins and peptides, as well as other thioethers, are also capable of reducing gold(III) to gold(I).[14] Even disulfide bonds react rapidly to reduce gold(III).[15,17] Thus, it appears that the bulk of the gold present *in vivo* is likely to be gold(I). Nonetheless, various bio-inorganic chemists have always been careful to point out the *potential* for oxidizing gold(I) to gold(III) *in vivo*,[7-9] and Smith and Brown actually predicted it.[9]

Gleichmann and co-workers have observed that gold drugs can be activated *in vivo* to a gold(III) metabolite that is responsible for some of the immunological side effects observed in chrysotherapy.[77,78] This finding is based on the observation that after mice have been tested with AuSTm (gold sodium thiomalate, Figure 1) for several weeks, gold(III) elicits a response in the popliteal lymph node assay (PLNA), but AuSTm does not. The PLN assay is important because it discriminates between the effects of a drug and those of its metabolites, in order to determine which is immunogenic. Subsequent research in another laboratory confirmed that T cells from human chrysotherapy patients are sensitized against gold(III) but not gold(I).[79] Thus, the formation of gold(III) *in vivo* is not only feasible, but demonstrably relevant to human therapy.

A cursory report by Beverly and Couri suggested that hypochlorous acid (HOCl), which is generated by the enzyme myleoperoxidase during the oxidative burst, can oxidize the gold in AuSTm to Au(III).[80] This finding has been extended to additional gold compounds.[74] For example, AuSTm, auranofin and $Au(CN)_2^-$ are oxidized to Au(III), and the ligands are oxidized preceding or in concert with the gold oxidation:[74]

$$AuSR \ + \ 4\,OCl^- \ \longrightarrow \ AuCl_4^- \ + \ RSO_3^- \tag{27}$$

$$Et_3PAuSATg \ + \ 5\,OCl^- \ \longrightarrow \ AuCl_4^- \ + \ ATgSO_3^- \ + \ Et_3P{=}O \tag{28}$$

This clearly establishes the chemical feasibility of gold oxidation by hypochlorite and provides a plausible mechanism for the findings of Gleichmann[77,78] and Verwilghen.[79]

An explanation for the apparent dichotomy between the observations that gold is present primarily as gold(I) *in vivo* and that T cells are sensitized to gold(III), not to the gold drugs themselves, is a redox cycle:[74]

$$\text{Drugs} \rightarrow \text{Protein-Au}^{I}\text{-Ligand} \qquad\qquad Au^{III}X_4^- \tag{29}$$

$$\overbrace{\qquad\qquad}^{OCl^-\ \text{(oxidative burst)}}$$

$$\underbrace{\qquad\qquad}_{\text{thiols, thioethers, disulfides}}$$

The operation of such a cycle is consistent with observations that while relatively low concentrations of gold are present during chrysotherapy ($10–50\,\mu M$ Au), the changes in tissue levels of metals, thiols, proteins, *etc.* in responding patients are much larger than can be accounted for on a stoichiometric basis.

Antigenic Peptides

Immunologists have found that the treatment of mice[77] and men[79] with anti-arthritic gold drugs generates T-cells that react to gold(III) but not to the original gold(I) drugs. In another context, industrial workers exposed to platinum, palladium, nickel or other metals frequently develop allergies and other immune reactions.[81,82] It is unlikely that the immune system is reacting directly to the metal ions, as they are too small to be recognized. Instead, there are a number of ways that metal ions can alter the formation and presentation of peptides to the immune system as it attempts to distinguish foreign matter from self.[81–83]

When macrophages and other antigen presenting cells digest foreign substances and broken-down self tissues, they do it by hydrolyzing proteins into short peptides and, ultimately, into individual amino acids. Certain of these peptides bind to large macromolecular structures called major histocompatibility complexes (MHC) or human leucocyte antigens (HLA). These MHCs are transferred to the outer surface of the macrophage cell membrane where T-cell receptors screen the MHC–peptide complexes (pathway 1 of the Scheme) and react with those carrying peptides of foreign origin. In that case the T-cells stimulate an immune response. Although peptides derived from self proteins do not normally react with T-cells in this way, certain auto-immune diseases arise when T-cells react to self proteins, which can occur due to a variety of reasons. Sometimes aberrant T-cells learn to recognize dominant self-peptides that are usually ignored; another cause is presentation of unfamiliar self peptides – designated cryptic – which stimulate a response. Among these autoimmune diseases is rheumatoid arthritis.

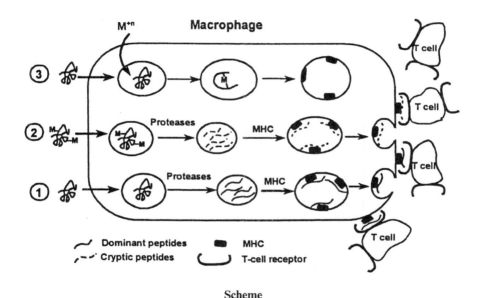

Scheme

Interestingly, immunological side effects to the gold drugs are common among those patients who respond successfully to chrysotherapy.[79] As a result, chrysotherapy was described by the Gleichmann group in Düsseldorf, Germany as a double-edged sword since it halts the autoimmune attack on the joints and simultaneously induces unwanted side effects. They have developed two bioinorganic model systems for studying the effects of gold on the immune system.

The first model is based on the ability of Au(III) to alter the response of T-cells to ribonuclease-A (RNase).[78,84,85] RNase injected into mice normally generates, *via* pathway 1, the peptide composed of amino acids 74–88 in the protein sequence, which is considered to be immunodominant and stimulates a T-cell response.

$RNase_{74-88}$:   H$_2$N-Gln-Ser-Tyr-Ser-Thr-Met-Ser-Ile-Thr-Asp-Cys-Arg-Glu-
Thr-Gly-COOH

If, however, the RNase is first pretreated with $AuCl_4^-$, peptides 7–21 and 94–108, which are considered to be cryptic since they are not normally recognized are generated (pathway 2) and stimulate a response from other T-cells:

RNase$_{7-21}$: H_2N-Lys-Phe-Glu-Arg-Gln-His-Met-Asp-Ser-Ser-Thr-Ser-Ala-
                     Ala-Ser-COOH

RNase$_{94-108}$: H_2N-Asn-Cys-Ala-Tyr-Lys-Thr-Thr-Gln-Ala-Asn-Lys-His-Ile-
                     Ile-Val-COOH

Treatment with $AuCl_4^-$ oxidizes the methionine residues ($CH_2CH_2SCH_3$) of the RNase[16] and, thereby, subtly alters the protein structure leading to the formation of the cryptic peptides. Thus, this system models the side effects observed in humans and mice where the secondary response to gold is not to the original drug, gold(I) thiomalate, but to a protein altered by $AuCl_4^-$. The second model system, also involving T-cells from mice, demonstrates that the effects of bovine insulin on T-cell response can be moderated by simultaneous exposure to gold(I) thiomalate.[86] The T-cells react to peptide A1–14 (the first 14 amino acids of the insulin A chain), which contains three cysteine residues located at positions 6, 7 and 11 (designated Ins$_{A1-14}$/CCC).

$$\overset{\displaystyle \ulcorner SH \ulcorner SH \qquad\qquad \ulcorner SH}{H_2N\text{-Gly-Ile-Val-Glu-Gln-Cys}_6\text{-Cys}_7\text{-Ala-Ser-Val-Cys}_{11}\text{-Ser-Val-Tyr-COOH}}$$

As discussed elsewhere in this chapter, the cysteine residues of serum albumin, hemoglobin and metallothionein show a high affinity for Au(I). Recent studies of a mutant (Ins$_{A1-14}$/SCC), where cysteine 6 is replaced by a serine which effectively changes the cysteine SH group to an OH, show that it can react with $Au(STm)_2^{5-}$ to form complexes with one or two Au(I) ions (A. Muñoz and C. F. Shaw III, unpublished results). The 1:1 complex is formed with loss of the thiomalate ligands, Au-Ins$_{A1-14}$/SCC:

$$\overset{\displaystyle \ulcorner\!\!\!-\!\!-\text{S-Au-S}\!\!-\!\!\!\urcorner}{H_2N\text{-Gly-Ile-Val-Glu-Gln-Ser-Cys}_7\text{-Ala-Ser-Val-Cys}_{11}\text{-Ser-Val-Tyr-COOH}}$$

When excess $Au(STm)_2^{5-}$ is used, a digold complex which retains two thiomalates is formed, $(TmSAu)2$-Ins$_{A1-14}$/SCC:

$$\overset{\displaystyle \ulcorner S\text{-Au-STm}^{3-} \qquad \ulcorner S\text{-Au-STm}^{3-}}{H_2N\text{-Gly-Ile-Val-Glu-Gln-Ser-Cys}_7\text{-Ala-Ser-Val-Cys}_{11}\text{-Ser-Val-Tyr-COOH}}$$

These structures are consistent with the tendency of gold to form two-coordinate linear structures with thiolate ligands. The gold adducts will have very different electronic and steric properties. In particular the steric constraints of the mono-gold complex will destroy the extended structure necessary for binding of the peptide to the MHC complex (pathway 3). Hence, the effect of the gold(I) thiomalate in this immune model system is presumably to alter the peptide structure and prevent its binding to the MHC molecule. This, in turn, suggests that the mechanism by which gold therapy causes remission of rheumatoid arthritis might be to prevent the presentation of a cysteine-containing immunogenic self-peptide to which the patients' T-cells react.

4 Anti-Tumour Activity

Interest over the last several decades in the potential anti-tumour activity of gold complexes has been driven by three common rationales: (1) analogy to the immunomodulatory properties underlying the benefit from gold(I) complexes in treating rheumatoid arthritis; (2) the structural analogy of square-planar gold(III) to platinum(II) complexes, which are potent anti-tumour agents; and (3) complexation of gold(I) or gold(III) with other active anti-tumour agents in order to enhance the activity and/or alter the biological distribution of the latter. Three reviews of the topic in various contexts have been published over the last decade, by Haiduc and Silvestru,[87] Sadler et al.[88] and Shaw.[89]

Auranofin and Analogues

Lorber and co-workers reported in 1979 that auranofin (Figure 1) inhibited the proliferation of HeLa cells in culture.[90] The uptake and incorporation of [³H] thymidine into DNA is reduced by brief exposure of cultured HeLa cells[91] or EBB-transformed lymphocytes[92] to $50\text{--}100\,\mu g\,dl^{-1}$ auranofin. The similarity of the irreversible inhibition of DNA synthesis by auranofin and by the clinically used cisplatin agent was noted.[90,92] When tested against P388 leukemia in mice, auranofin produced significant, dose-dependent increases in life span.[91,93] Further testing of related complexes (Table 4) established that the tertiaryphosphine gold(I) complexes with thiosugar ligands (R3PAuS-sugar) had maximal anti-tumour activity in vivo when tested against P388 leukemia. Analogues with thiolate ligands, halide ions or other donor atoms replacing the thiosugar are less potent. Oligomeric thiolates, $[AuSR]_n$, including the injectable anti-arthritic drugs myochrysine and solganol, have very little in vivo activity.[93,94] Thus, the general trend in activity is

$$R_3PAuS\text{-sugar} > R_3PAuS\text{-alkyl} \approx R_3PAuX > [AuSR]_n$$

Cytotoxicity is a necessary, but not sufficient criterion for anti-tumour activity. All of the compounds with strong anti-tumour activity in vivo also have potent cytotoxicity when measured against B16 melanoma in vitro. The converse relationship was not observed, as some highly cytotoxic complexes have limited in vivo efficacy. This relationship demonstrates the importance of in vivo metabolism and animal testing in drug design and development.

Unfortunately, auranofin is not a broad spectrum anti-tumour agent. Its activity in vivo is limited to the P388 cell line (Table 5) when administered intraperitoneally (i.p.).[35]

Table 4 *Activity of gold complexes against P388 leukemia and B16 melanoma cells*

Complex	B16 in vitro[a] $I_0(\mu M)$	P388 in vivo[b] $ILS_{max}(\%)$	Reference
Auranofin analogues			
$Et_3PAuS-\beta$-Glu(Ac)$_4$ [auranofin]	1.5	70	82
$Et_3PAuS-\beta$-Glu	2	68	82
$Et_3PAuS-\beta$-Glu(CONHMe)$_4$	7	58	82
$Et_3PAuS-\alpha$-Glu(Ac)$_4$ [epi-auranofin]	4	65	93
$Et_3PAuS-\beta$-Gal(Ac)$_4$	4	88	93
Et_3PAuS-Glutathione	2	32	93
$Et_3PAuSCN$	1	36	93
$Et_3PAuSCH_3$	60	36	93
Et_3PAuCl	1	36	93
Et_3PAuCN	0.4	68	93
Et_3PAuCH_3	1	55	93
$Et_3PAuPEt_3^+Cl^-$	1	36	93
Oligomeric gold(I) thiolates			
$[AuSTm]_n$	60	24	93
$[AuSGlu]_n$	166	15	93
$[AuSGlu(Ac)_4]_n$	150	14	93
DPPE complexes			
$DPPE(Au^ICl)_2$	8	—	93
$DPPE(Au^{III}Cl_3)_2$	15	—	93
$DPPE(Au^ISGlu)2$	4	—	96
$[Au(DPPE)_2^+]Cl$	2	—	100, 101
DPPE	60	—	93
$DPPE-O_2$	inactive	—	93

Bis(DPPE)gold(I) analogues $[Au(R_2PYPR'_2)_2]X$

R,R'	Y	X			
C_6H_5	$(CH_2)_2$	Cl	4.5	83 ± 25	100, 101
C_6H_5	$(CH_2)_2$	Br	—	70, 83	100, 101
C_6H_5	$(CH_2)_2$	NO_3	4	90 ± 17	100, 101
C_6H_5	$(CH_2)_3$	Cl	0.6	89 ± 28	100, 101
C_6H_5	$CH_2=$	Cl	2	92 ± 26	100, 101
3-F-C6H$_4$	$(CH_2)_2$	Cl	—	45.55	100, 101
2-C$_2$H$_5$N	$(CH_2)_2$	Cl	—	75 ± 4	100, 101
C_6H_5,C_2H_5	$(CH_2)_2$	Cl	5	54 ± 16	100, 101
C_2H_5	$(CH_2)_3$	Cl	17	40, 30	100, 101
cisplatin			—	125 ± 38	

[a]B16 melanoma tested *in vitro* (IC_{50} is the dose that inhibits growth by 50%). [b]*In vivo*, using B6D2F1 mice injected intraperitoneally with the maximum tolerated dose (MTD) daily for 5 days.

Table 5 *Activity of selected gold complexes against various tumour lines*

Tumour (implantation)	Auranofin Ref. 35	DPPE(AuCl)₂ Ref. 98	[Au(DPPE)₂]Cl Ref. 100
P388 leukemia (i.p.)	active	active	active
P388 leukemia (i.v.)	inactive	inactive	inactive
M5076 reticulum cell sarcoma (i.p.)	inactive	active	active
B16 Melanoma (i.p.)	inactive	active	active
colon carcinoma 26 (i.p.; s.c.)	inactive	—	inactive
Madison 109 lung carcinoma (i.p.; s.c.)	inactive	—	—
Lewis lung carcinoma (i.p.; s.c.)	inactive	—	—
mammary adenocarcinoma 16/c (s.c.)	inactive	active	active
mammary adenocarcinoma 13/c (s.c.)	inactive	—	inactive
L1210 leukemia (i.p.)	—	active	active

Stocco *et al.* found that $(Ph_3PAu)_2(\mu\text{-DTE})$, where DTE = dithioerythritol, is active (T/C = 167) against Ehrlich ascites tumours in mice.[94] Triphenylphosphine(8-thiotheophyllinato)gold(I), $Ph_3PAutTP$ (Figure 7) is active against several cell types.[95] The rationale for using the tTP ligand, a modified purine base, is the possibility of increased DNA–gold interactions. The IC_{50} values are 0.6 μM when tested against normal FLC cells and 1.8 μM against doxorubicin-resistant cells (Dox-RFLC), which carry a multi-drug resistance gene and differ significantly from the normal FLC cells in membrane fluidity.

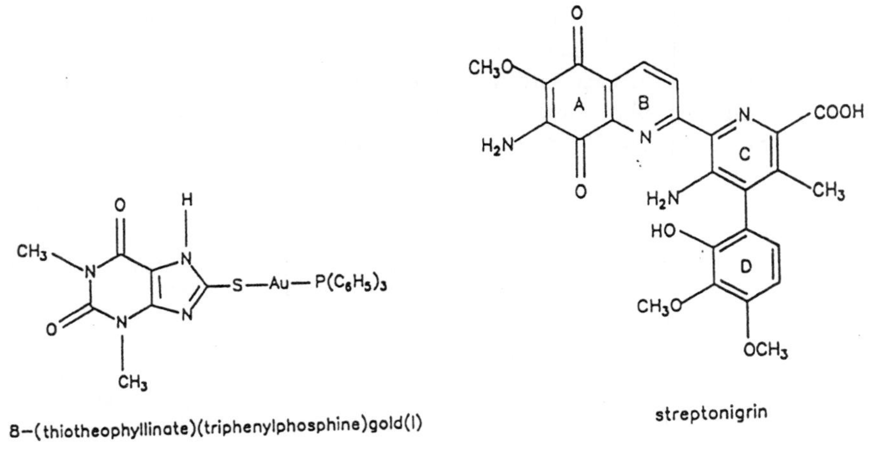

8-(thiotheophyllinate)(triphenylphosphine)gold(I)

streptonigrin

Figure 7 *The structures of Ph₃PAutTP and streptonigrin*

Bis(diphenylphosphine)ethane Gold Complexes

As an outgrowth of the work on auranofin analogues, digold complexes with a bridging bis(diphenylphosphine)ethane (DPPE) ligand (Figure 8), were investigated for anti-tumour activity.[96,97] Table 4 shows that the ligand itself has limited activity, and its oxidation product (DPPE=O_2) has no activity, while the digold complexes are considerably more potent.[97] (μ-DPPE)(AuSGlu)$_2$ inhibits DNA biosynthesis [measured as (^{3}H)-thymidine incorporation into DNA] more effectively than it inhibits RNA or protein biosynthesis (measured as [^{3}H]-uridine and [^{3}H]leucine incorporation into RNA or proteins, respectively).[96] Comparisons of DPPE and (μ-DPPE)(AuSGlu)$_2$ established that the gold complex is a better inhibitor of DNA polymerase α both *in vitro* and in permeabilized KB cells.[97] When tested against other cell lines, (μ-DPPE)(AuCl)$_2$ showed much broader activity than did the auranofin analogues discussed above (Table 5).[98] The activity is not unique to gold complexes, however,[97] as Cu(II) also enhances the cytotoxicity of DPPE in the B16 melanoma assay and increases its ability to inhibit DNA polymerase α. The role of copper and gold in the complexes may be to stabilize DPPE against oxidation.[97]

In culture medium (and presumably *in vivo*), the digold complexes rearrange to form a new species, which is more active than the precursors:[99]

$$(\mu\text{-DPPE})(\text{AuX})_2 \rightarrow \text{Au(DPPE)}_2{}^+ \tag{30}$$

The reaction product, Au(DPPE)$_2{}^+$ (Figure 7), has a useful spectrum of activity against various tumours (Table 1).[100,101] Against B16 melanoma *in vitro*, the IC$_{50}$ value was 2 μM (Table 3), slightly more potent than the open chain digold compounds of DPPE. It inhibits DNA, RNA and protein synthesis, but the greatest effect is on protein synthesis, which can be shut down by 90 minutes exposure to 15 μM Au(DPPE)$_2{}^+$. In contrast, auranofin and its analogues exert their primary effect on DNA synthesis.[35,93]

Au(DPPE)$_2{}^+$ is stable in serum for 25 hours, and in the presence of glutathione for 8 days.[101] The stability and inertness result from two factors, the chelate effect (since DPPE coordinates *via* two phosphorus donor atoms) and the greater stability of phosphine over thiolate ligands. The complex reacts with Cu(II), forming Au$_2$(DPPE)$_2{}^{2+}$ and a precipitate containing Cu and DPPE.[101]

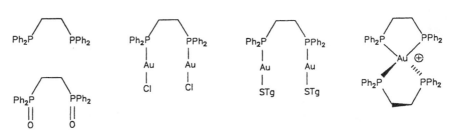

Figure 8 *DPPE, DPPE–O$_2$, and three DPPE–gold complexes with antitumour activity*

Structure–function relationships were examined for complexes with a variety of related diphosphine ligands.[101] Generally the phenylphosphine complexes are more active than are ethyl, pyridyl or fluorophenyl derivatives.[101] As would be expected when a cationic complex is the active species, the nature of the anion (e.g. Cl^-, Br^- or NO_3^-) does not substantially affect the potency of the drug.

Snyder et al.[97] showed that the diphos analogues that have enhanced cytotoxicity when bound to copper are active in vivo against P388, whereas the analogues that are not more cytotoxic in the presence of copper are inactive. The latter group includes ligands that are structurally unable to chelate to a metal ion because of rigidity in the linker region:

$$Ph_2P-C\equiv C-PPh_2, \; trans\text{-}Ph_2P-C=C-PPh_2, \; and \; para\text{-}Ph_2P-C_6H_4-PPh_2.$$

Substituting Ph_2As for Ph_2P also eliminates the Cu-induced enhancement and the in vivo activity,[97] but not in vitro activity against several cell lines.[102] From these results, Snyder et al. suggested that Cu(I) may function to transport DPPE, or, conversely, that DPPE may transport Cu(I) to the target site, or that the entire Cu-DPPE molecule is necessary for activity.[97]

Gold Complexes of Known Anti-Tumour Agents

A common rationale for the design of potential gold-based, anti-tumour agents is to attach gold(I) or gold(III) to a compound that has anti-tumour potency and good metal-ligating donor atoms. For example, Ph_3PAu-nucleotide complexes of 2-thiouracil, 5-fluorouracil, 5-fluorodeoxyuridine, thymidine and 6-mercaptopurine were investigated.[103] Of these, the thymidine complex produced a T/C (%) of 171 against P388 tumours in mice treated with 43 $\mu mol\,kg^{-1}$, 30 $mg\,kg^{-1}$ on days 1, 3, and 5 after tumour implantation. Ph_3PAu-5-fluorodeoxyuridine and Ph_3PAu-tegafur [tegafur = 5'-fluoro-1-(tetrahydro-2-furanyl)-2,4-($1H,3H$)-pyrimidinedionato-N^3, a known anti-neoplastic agent] have been characterized and tested against L1210 leukemia in mice: the activity of the gold–5-fluorouridine complex was greater than that of the ligand itself, but for the tegafur complex there was no increase.[104]

Streptonigrin (Figure 7) is a substituted 7-amino-quinoline-5,8-dione with anti-tumour activity that is complicated by high toxicity. It forms a highly stable gold(III) complex[105] that is not altered by exposure to serum albumin, a high affinity binding site for gold(I), and a good reducing agent for gold(III). However, the activity of the complex against P388 tumour cells in vitro is the same as that of the streptonigrin itself, $IC_{50} = 0.5\,\mu g\,ml^{-1}$.[105]

Cisplatin Analogues

As cisplatin [$(H_3N)_2PtCl_2$] developed into a clinically important chemotherapy agent, many inorganic chemists were attracted to the structural similarities of gold(III) and platinum(II). Both are d^8 metal ions and form square-planar, four-coordinate structures. Sadler's review of anti-tumour activity by compounds of

Group IB (Cu, Ag, Au) elaborates this rationale.[88] Two dimethylgold(III) complexes noteworthy for their similarity to cisplatin, $[Me_2AuCl_2][AsPh_4]$ and $Me_2Au(\mu\text{-SCN})_2AuMe_2$, had activities exceeding 120%. However, a variety of gold(III) complexes without the redox-stabilizing methyl groups were inactive: $PhAuCl_2Y$ (Y = pyridine, Bu_2S); $AuCl_4^-$; $pyAuCl_3$ and $[Au(en)_2]Cl_3$.[88] While the rationale based on coordination geometries is attractive to structural inorganic chemists, it fails *in vitro*, where the greater oxidation potentials and more facile ligand exchange reactions of gold(III) lead to rapid, irreversible formation of new and inactive metabolites that cannot cross-link DNA with the same efficiency at Pt(II).

Prospects for Gold-Based Anti-Cancer Drugs

The intense testing of new gold complexes for anti-cancer activity in the 1980s was stimulated by the finding that auranofin had anti-tumour and cytotoxic properties.[90–92] The center of attention quickly passed from auranofin analogues to $(\mu\text{-DPPE})Au_2X_2$ digold complexes, to $Au(DPPE)_2^+$ and its copper analogue. Unfortunately, the finding of significant cardiotoxicity in animal testing has brought this line of research to a sudden halt.[106]

Despite the absence of any clinical trials of a gold complex to date, the potential for developing new cytotoxic gold complexes that have anti-tumour activity is clear. This is demonstrated, for example, by the broad-spectrum activity of $Au(DPPE)_2^+$. The lability of gold(I) and gold(III) to ligand exchange and of gold(III) to redox reactions must be recognized in efforts to synthesize new drugs. Robust new ligand structures that can move gold through cell membranes and into the cytoplasm and perhaps into the nucleus are required. The chemistry and biology of the DPPE complexes of copper(I) and gold(I) and the streptonigrin complex of gold(III) demonstrates that the chelate effect shows promise for developing compounds that will remain intact for longer periods *in vivo* and permit the delivery of gold to potential target sites. The modification of known anti-tumour agents also deserves further exploration. Whether, in fact, the DPPE or the metal ions were the active component of the gold and copper complexes, the striking activity of the complexes shows the untapped potential of inorganic species in general and gold compounds in particular as medicinal agents.

5   Anti-HIV Activity

Investigations of the anti-HIV activity of gold complexes were stimulated by reports that AuSTg inhibits reverse transcriptase (RTase), an enzyme that converts viral RNA into DNA in the host cell.[107] AuSTg is indeed active in the cell-free extracts where it was studied, but it is unable to enter the cells where RTase acts. A different mechanism of action has been proposed for $Au(STg)_2^-$, which can be generated *in situ* from AuSTg and TgSH.[108] $Au(STg)_2^-$ inhibits the infection of MT-4 cells by HIV strain HL4-3 without inhibiting the RTase activity in the intact virons. The critical target site has been tentatively identified as a thiol group, cys-532 on gp160, which is a glycoprotein of the viral envelope.[97] Interestingly, $Au(STm)_2{}^{5-}$ is also active, but the oligomeric 1:1 thiolates, AuSTm and AuSTg, are not active. Auranofin and two analogues ($Et_3PAuSTg$ and Et_3PAuCl), which are able to enter cells, are inactive against the HL4-3 strain (Table 6). Several trialkylphosphinegold(I) cyanides (R_3PAuCN) are inactive below the onset of cytotoxicity to the host T cells at $\sim 1\,\mu M$ concentrations (Table 6).[109]

Tepperman and colleagues[73] have found that $Au(CN)_2{}^-$ is taken up into H9 cells, a continuous T-cell line that is susceptible to HIV infection. At concentrations as low as 20 ppb, it retards the proliferation of HIV in these cells. The concentration used is well tolerated in arthritis patients, which suggests that $Au(CN)_2{}^-$ may have promise as a complement for existing HIV treatments.[73]

6 Conclusions

Our understanding of gold complexes at the biochemical and cellular levels has evolved rapidly in the last decade. Among the most exciting findings are the roles of $Au(CN)_2{}^-$, gold(III), and protein–gold complexes in the metabolism of gold compounds during chrysotherapy. Among the most important conclusions is that the medicinal agents are actually pro-drugs. These findings are consistent with an earlier prediction that the bioinorganic chemistry of gold complexes requires the formation of common metabolites from the parenteral drugs and auranofin.[24] Results from numerous laboratories spanning a range of disciplines, as presented here, are consistent with this view. Similar conclusions have been reached from consideration of the pharmacology of gold complexes.[110] Despite the great gains in understanding cellular and biochemical aspects of gold metabolism in the last decade, there has been as little real progress in understanding the mechanisms of chrysotherapy as in the preceding half century. The bioinorganic

Table 6 *Toxicity and HIV-1 antiviral activity of gold phosphines*[a]

	MTT survival results (%)[b]					
	Toxicity (Au only)			*+ HIV-1(NL43)*		
Concentration (μM)	1.0	0.10	0.010	1.0	0.10	0.010
$Et_3PAuSATg$	61	97	122	0	7	4
$Et_3PAuSTg$	130	108	96	7	8	0
Et_3PAuCl	92	86	96	1	7	0
Me_3PAuCN	0	88	73	0	9	8
Et_3PAuCN	0	78	86	0	8	8
iPr_3PAuCN	0	100	103	0	6	0
Ph_3PAuCN	0	101	91	0	8	7
Cy_3PAuCN	60	85	127	2	6	9
$KAu(CN)_2$	0	71	59	0	6	6

[a]Unpublished data of T. Okada, M. Gurney, A. L. Arendt and C. F. Shaw III. [b]MTT Survival Assay using MT-4 cells incubated at 37 °C for 5 days (the assay is described in Ref. 108); values are % surviving T-cells compared with controls.

chemistry clearly shows that the medicinal agents are *pro-drugs* with short half-lives *in vivo*. There is a compelling need for new experimental designs involving the metabolites, because they either (1) are the active species or (2) generate it at the presently unidentified sites of action.

The possibility of developing anti-tumour agents based on gold or using gold to modulate organic drugs through ligation is a tantalizing prospect. The recent reports of anti-HIV activity for cyanide and thioglucose derivatives may be even more significant and clearly warrant active pursuit of their mechanism(s) of action.

References

1 W.F. Kean, C.J.L. Lock and K. Howard-Lock, *Inflammopharmacology*, 1991, **1**, 103.
2 M.C. Grootveld, M.T. Razi and P.J. Sadler, *Clin. Rheumatol.*, 1984, **3**, 5.
3 G.J. Higby, *Gold Bull.*, 1982, **15**, 130.
4 D.T. Felson, J.J.Anderson and R.F. Menan, *Arthr. Rheum.*, 1990, **33**, 1449.
5 I. Jaffe, in *The Columbia University Complete Home Medical Guide*, ed. D.F. Tapley, T.Q. Morris, L.P. Rowland and R.J. Wiess, Crown, New York, 1989, pp. 602–625.
6 R.C. Elder, Z. Zhao, Y. Zhang, J.G. Dorsey, E.V. Hess and K. Tepperman, *J. Rheumatol.*, 1993, **20**, 268.
7 P.J. Sadler, *Struct. Bonding*, 1976, **29**, 171.
8 C.F. Shaw III, *Inorg. Persp. Biol. Med.*, 1979, **2**, 278.
9 D.H. Brown and W.E. Smith, *Chem. Soc. Rev.*, 1980, **9**, 217.
10 R.J. Puddephatt, *The Chemistry of Gold*, Elsevier, Amsterdam, 1978.
11 P.G. Jones, *Gold Bull.*, 1981, **14**, 102, 159; 1983, **16**, 114.
12 M. Melnick and R.V. Parish, *Coord. Chem. Rev.*, 1986, **70**, 157.
13 H. Schmidbaur, *Gold Bull.*, 1990, **23**, 11.
14 M.C. Gimeno and A. Laguna, *Chem. Rev.*, 1997, **97**, 511.
15 P.L. Witkiewicz and C.F. Shaw III, *J.C.S. Chem. Commun.*, 1981, 1111.
16 A.A. Isab and P.J. Sadler, *Biochim. Biophys. Acta*, 1977, **492**, 322.
17 C.F. Shaw III, M.P. Cancro, P.L. Witkiewicz and J. Eldridge, *Inorg. Chem.*, 1980, **19**, 3198.
18 C.F. Shaw III, ed., 'Proceedings of the 3rd International Conference on Gold and Silver in Medicine', *Metal Based Drugs*, 1994, **1**, 350–529.
19 R.V. Parish and S.M. Cottrill, *Gold Bull.*, 1987, **20**, 3.
20 W.E. Smith and J. Reglinski, *Persp. Bioinorg. Chem.*, 1991, **1**, 183.
21 M.C. Grootveld, M.T. Razi and P.J. Sadler, *Clin. Rheumatol.*, 1983, **3** Suppl. 1, 5.
22 S.T. Crooke, R.M. Snyder, T.R. Butt, D.J. Ecker, H.S. Allaudeen, B. Monia and C.K. Mirabelli, *Biochem. Pharmacol.*, 1986, **35**, 3423.
23 R.C. Elder and M.K. Eidsness, *Chem. Rev.*, 1987, **87**, 1027.
24 C.F. Shaw III, *Comments Inorg. Chem.*, 1989, **8**, 233.
25 C.F. Shaw III and M.M. Savas, in *Metallothioneins: Synthesis, Structure and Properties of Metallothioneins, Phytochelatins and Metal-Thiolate Complexes*, ed. M.J. Stillman, C.F. Shaw III and K.T. Suzuki, VCH Publishers, New York, 1992, pp. 144–163.
26 S.R. Rudge *et al.*, *J. Rheumatol.*, 1984, **11**, 150.
27 H.A. Schwartz *et al.*, *Am. J. Physiol.*, 1960, **199**, 76.
28 A.P. Intoccia *et al.*, in *Bioinorganic Chemistry of Gold Coordination Compounds*, ed. B.M. Sutton and R.G. Franz, SK&F, Philadelphia, 1983, pp. 21.

29 S.M. Cottrill, H.L. Sharma, D.B. Dyson, R.V. Parish and C.A. McCauliffe, *J. Chem Soc. Perkin Trans.* 2, 1989, 53.

30 W.E. Smith and J. Reglinski, *Persp. Bioinorg. Chem.*, 1991, **1**, 183.

31 W.E. Smith, J. Reglinski, S. Hoey, D.H. Brown and R.D. Sturrock, *Inorg. Chem.*, 1990, **29**, 5190.

32 R.M. Snyder, C. Mirabelli and S.J. Crooke, *Biochem. Pharmacol.*, 1986, **35**, 923.

33 C.F. Shaw III, A.A. Isab, M.T. Coffer and C.K. Mirabelli, *Biochem. Pharmacol.*, 1990, **40**, 1227.

34 R.M. Snyder, C.K. Mirabelli and S.T. Crooke, *Biochem. Pharmacol.*, 1986, **35**, 923.

35 C.K. Mirabelli, R.K. Johnson, C.M. Sung, L. Faucette, K. Muirhead and S.T. Crooke, *Cancer Res.*, 1985, **45**, 32.

36 M.P. Arizti, A. Garcia-Orad, F. Sommer, L. Silvestro, P. Massiot, P. Chevallier, J.M. Gutierrez-Zorrilla, E. Colacio, M. Martinez de Pancorbo and H. Tapiero, *Anticancer Res.*, 1991, **11**, 625.

37 M.N. Akhtar, A.A. Isab and A.R. Al-Arfaj, *J. Inorg. Biochem.*, 1997, **66**, 197.

38 D. Carter and J.X. Ho, *Adv. Protein Chem.*, 1994, **45**, 153.

39 C.F. Shaw III, N.A. Schaeffer, R.C. Elder, M.K. Eidsness, J.M. Trooster and G.H.M. Calis, *J. Am. Chem. Soc.*, 1984, **106**, 3511.

40 M.T. Coffer, C.F. Shaw III, M.K. Eidsness, J.W. Watkins II and R.C. Elder, *Inorg. Chem.*, 1986, **25**, 333.

41 D.J. Ecker, J.C. Hempel, B.M. Sutton, R. Kirsch and S.T. Crooke, *Inorg. Chem.*, 1987, **26**, 3139.

42 M.T. Razi, G. Otiko and P.J. Sadler, *Am. Chem. Soc. Symp. Ser.*, 1983, **209**, 371.

43 N.A. Malik, G. Otiko and P.J. Sadler, *J. Inorg. Biochem.*, 1980, **12**, 317.

44 J. Roberts, J. Xiao, B. Schliesman, D.J. Parsons and C.F. Shaw III, submitted for publication.

45 J. Christodoulou, P.J. Sadler and A. Tucker, *Eur. J. Biochem.*, 1994, **225**, 363.

46 O.M. Ni Dhubhghaill, P.J. Sadler and A. Tucker, *J. Am. Chem. Soc.*, 1992, **114**, 1117.

47 M.T. Coffer, C.F. Shaw III, A.L. Hormann, C.K. Mirabelli and S.T. Crooke, *J. Inorg. Biochem.*, 1987, **30**, 177.

48 A.A. Isab, C.F. Shaw III and J. Locke, *Inorg. Chem.*, 1988, **27**, 3406.

49 C.F. Shaw III, A.A. Isab, J.D. Hoeschele, M. Starich, J. Locke, P. Schulteis and J. Xiao, *J. Am. Chem. Soc.*, 1994, **116**, 2254.

50 M. Starich and C.F. Shaw III, unpublished observations.

51 A.A. Isab, A.L. Hormann, M.T. Coffer and C.F. Shaw III, *J. Am. Chem. Soc.*, 1988, **110**, 3278.

52 A.A. Isab, D.T. Hill and C.F. Shaw III, unpublished observations.

53 C.F. Shaw III, M.T. Coffer, J. Klingbeil and C.K. Mirabelli, *J. Am. Chem. Soc.*, 1988, **110**, 729.

54 M.J. Stillman, C.F. Shaw III and K.T. Suzuki, eds., *Metallothioneins: Synthesis, Structure and Properties of Metallothioneins, Phytochelatins and Metal Thiolate Complexes*, VCH Publishers, New York, 1992, 443 pp.

55 A. Bakka, L. Endresen, A.B.S. Johnsen, P.D. Edminson and H.E. Rugstad, *Toxicol. Appl. Pharm.*, 1981, **61**, 215.

56 L. Endresen, A. Bakka and H.E. Rugstad, *Cancer Res.*, 1983, **43**, 2918.

57 G. Schmitz, D.T. Minkel, D. Gingrich and C.F. Shaw III, *J. Inorg. Biochem.*, 1980, **12**, 293.

58 E.M. Mogilnicka and M. Webb, *Biochem. Pharmacol.*, 1983, **32**, 1341.

59 J.E. Laib, C.F. Shaw III, D.H. Petering, M.K. Eidsness, R.C. Elder and J.S. Garvey, *Biochemistry*, 1985, **24**, 1977.

60 M.M. Savas, D.H. Petering and C.F. Shaw III, *Inorg. Chem.*, 1990, **29**, 403.

61 A.A. Muñoz and C.F. Shaw III, unpublished data.

62 C.F. Shaw III and J. Laib, *Inorg. Chim. Acta*, 1896, **123**, 197.

63 B.P. Monia, T.R. Butt, D.J. Ecker, C.K. Mirabelli and S.T. Crooke, *J. Biol. Chem.*, 1986, **261**, 10957.

64 J. Chaudiere and A.L. Tappel, *J. Inorg. Biochem.*, 1984, **20**, 313.

65 A.L. Tappel, *Cur. Topics Cell. Reg.*, 1984, **24**, 87.

66 J.R. Roberts and C.F. Shaw III, *Biochem. Pharmacol.*, 1998, **55**, 1.

67 D.W. James, N.W. Ludvigsen, L.G. Cleland and S.C. Milazzo, *J. Rheumatol.*, 1982, **9**, 532.

68 G.G. Graham, T.M. Haavisto, H.M. Jones and G.D. Champion, *Biochem. Pharmacol.*, 1984, **33**, 1257.

69 G.G. Graham, G.D. Champion and J.B. Ziegler, *Metal-Based Drugs*, 1994, **1**, 395.

70 R.C. Elder, W.B. Jones, Z. Zhao, J.G. Dorsey and K. Tepperman, *Metal-Based Drugs*, 1994, **1**, 363.

71 G. Lewis and C.F. Shaw III, *Inorg. Chem.*, 1986, **25**, 58.

72 G.G. Graham, J.R. Bales, M.C. Grootveld and P.J. Sadler, *J. Inorg. Biochem.*, 1985, **25**, 163.

73 K. Tepperman, Y. Zhang, P.W. Roy, R. Floyd, Z. Zhao, J.G. Dorsey and R.C. Elder, *Metal-Based Drugs*, 1994, **1**, 433.

74 C.F. Shaw III, S. Schraa, E. Gleichmann, Y.P. Grover, L. Dunemann and A. Jagarlamudi, *Metal-Based Drugs*, 1994, **1**, 351.

75 A. Canumalla, S. Schraa, A.A. Isab, C.F. Shaw III, E. Gleichmann, L. Dunemann and M. Turfeld. *J. Biol. Inorg. Chem.*, 1998, **3**, 1.

76 R.C. Elder, K.G. Tepperman, M.K. Eidsness, M.J. Heeg, C.F. Shaw III and N.A. Schaeffer, *ACS Symposium Ser.*, 1983, **209**, 385.

77 D. Schuhmann, M. Kubicka-Muranyi, J. Mirtcheva, J. Günther, P. Kind and E. Gleichmann, *J. Immunol.*, 1990, **145**, 2132.

78 K. Takahashi, P. Griem, C. Goebel, J. Gonzalez and E. Gleichmann, *Metal-Based Drugs*, 1994, **1**, 483.

79 J. Verwilghen, G.H. Kingsley, L. Gambling and G.S. Panayi, *Arthr. Rheum.*, 1992, **35**, 1413.

80 B. Beverly and D. Couri, *Fed. Proc.*, 1987, **46**, 854.

81 P. Griem and E. Gleichmann, *Curr. Opin. Immunol.*, 1995, **7**, 831.

82 P. Griem, E. Gleichmann and C.F. Shaw III in *Toxicology of the Immune System*, ed. D.A. Lawrence, Pergamon Press, New York, 1997, pp. 323–338.

83 C. Goebel, M. Kubicka-Muranyi, T. Tonn, J. Gonzalez and E. Gleichmann, *Arch. Toxicol.*, 1995, **64**, 450.

84 P. Griem, K. Panthel, H. Kalbacher and E. Gleichmann, *Eur. J. Immunol.*, 1996, **26**, 279.

85 P. Griem, C. von Vultée, K. Panthel, S.L. Best, P.J. Sadler and C.F. Shaw III, submitted, 1997.

86 P. Griem, K. Takahashi, H. Kalbacker and E. Gleichmann, *J. Immunol.*, 1995, 1575.

87 I. Haiduc and C. Silvestru, *In Vivo*, 1989, **3**, 285.

88 P.J. Sadler, M. Nasr and V.L. Narayanan, in *Platinum Coordination Complexes in Cancer Chemotherapy*, ed. M.P. Hacker, E.B. Douple and I.H. Krakoff, Martinus Nijhoff Publishing, Boston, 1984, pp. 290–304.

89 C.F. Shaw III, in *Metal Compounds in Cancer Therapy*, ed. S.P. Fricker, Chapman and Hall, London, 1994, pp. 46–64.

90 T.M. Simon, D.H. Kunishima, G.J. Vibert and A. Lorber, *Cancer*, 1979, **44**, 1965.

91 T.M. Simon, D.H. Kunishima, G.J. Vibert and A. Lorber, *Cancer Res.*, 1981, **41**, 94.
92 T.M. Simon, D.H. Kunishima, G.J. Vibert and A. Lorber, *J. Rheumatol. Suppl.* 5, 1979, 91.
93 C.K. Mirabelli, R.K. Johnson, D.T. Hill, L. Faucette, G.R. Girard, G.Y. Kuo, C.M. Sung and S.T. Crooke, *J. Med. Chem.*, 1986, **29**, 218.
94 C.F. Shaw III, A. Beery and G.C. Stocco, *Inorg. Chim. Acta*, 1986, **123**, 213.
95 M.M. De Pancorbo, A. Garcia-Orad, M.P. Arizti, J.M. Gutjierrez-Zorrilla and E. Colacia, in *Metal Ions in Biology and Medicine*, ed. P.H. Collery, L.A. Porier, M. Marfait and J.C. Eitenne, John Libbey, Eurotext, Paris, 1990, pp. 385–389.
96 C.K. Mirabelli, B.D. Jensen, M.R. Mattern, C.M. Sung, S.-M. Mong, D.T. Hill, S.W. Dean, P.S. Schein, R.K. Johnson and S.T. Crooke, *Anti-Cancer Drug Des.*, 1986, **1**, 223.
97 R.M. Snyder, C.K. Mirabelli, R.K. Johnson, C.M. Sung, L.F. Faucette, F.L. McCabe, J.P. Zimmerman, M. Whitman, J.C. Hempel and S.T. Crooke, *Cancer Res.*, 1986, **46**, 5054.
98 R.K. Johnson, C.K. Mirabelli, L.F. Faucette, F.L. McCabe, B.M. Sutton, D.L. Bryan, G.R. Girard and D.T. Hill, *Proc. Am. Assoc. Cancer Res.*, 1985, **26**, 254.
99 S.J. Berners-Price, P.S. Jarrett and P.J. Sadler, *Inorg. Chem.*, 1987, **26**, 3074.
100 S.J. Berners-Price, C.K. Mirabelli, R.K. Johnson, M.R. Mattern, F.L. McCabe, L.F. Faucette, C.-M. Sung, S.-M. Mong, P.J. Sadler and S.T. Crooke, *Cancer Res.*, 1986, **46**, 5486.
101 S.J. Berners-Price, G.R. Girard, D.T. Hill, B.M. Sutton, P.S. Jarrett, L.F. Faucette, R.K. Johnson, C.K. Mirabelli and P.J. Sadler, *J. Med. Chem.*, 1990, **33**, 1386.
102 O.M. Ni Dhubhghaill, P.J. Sadler and R. Kuroda, *J. Chem. Soc., Dalton Trans.*, 1990, 2913.
103 K.C. Agrawal, K.B. Bears, D. Marcus and H.B. Jonassen, *Proc. Am. Assoc. Cancer Res.*, 1978, **69**, 110.
104 T. Amagai, T.K. Miyamoto, H. Ichida and Y. Sasaki, *Bull. Chem. Soc. Jpn.*, 1989, **62**, 1078.
105 A. Moustatih and A. Garnier-Suillerot, *J. Med. Chem.*, 1989, **32**, 1426.
106 G.D. Hoke, R.A. Macia, P.C. Meunier, P.J. Bugelski, C.K. Mirabelli, G.F. Rush and W.D. Matthews, *Toxicol. Appl. Pharmacol.*, 1989, **100**, 293.
107 H. Blough, Abstracts 3rd International Conf. Gold and Silver in Medicine, 1990, Manchester, UK, p. 14.
108 T. Okada, B.K. Patterson, S.-Q. Ye and M.E. Gurney, *Virology*, 1993, **192**, 631.
109 T. Okada, M.E. Gurney, A.L. Arendt and C.F. Shaw III, unpublished data.
110 G.G. Graham, G.D. Champion and J.B. Ziegler, *Inflammopharmacology*, 1991, **1**, 99.

CHAPTER 4

Nitric Oxide in Physiology and Medicine

ANTHONY R. BUTLER[1] AND PETER RHODES[2]

[1]School of Chemistry, St. Andrews University, St. Andrews KY16 9ST, UK
[2]Department of Pharmacology, Ninewells Hospital, Dundee DD1 9SY, UK

1 Introduction

That nitric oxide (NO) has a number of important roles in animal physiology is now established beyond doubt. The first to be delineated was that in vasodilation. A key enzyme in arterial muscle relaxation (and hence artery dilation) is guanylate cyclase, and it had been known for some time that, *in vitro*, NO will activate the enzyme.[1] In 1987 it was demonstrated, as well, that NO is the endogenous molecule responsible for activation of guanylate cyclase.[2,3] NO is also generated in activated macrophages (cells that are an important part of the non-specific immune system) and leads to the formation of nitrite and nitrite ions.[4-6] The presence of the same process in the brain[7] and peripheral nerves[8] was soon established. The natural substrate for NO production was discovered to be arginine, and the process was shown to be effected by the enzyme nitric oxide synthase (NOS). The other product of the reaction is citrulline.[9] Details of the discovery and confirmation of the arginine-NO pathway in animal

physiology have been chronicled in a number of reviews[10] and will not be retold here. Rather we will concentrate on the consequences of these discoveries for the physiology, pathology, and therapy. More recently has absolute proof been obtained for production of NO by humans rather than other species. This work employed direct gas chromatography–mass spectrometry (GC–MS) detection of NO in the exhalate of normal subjects.[11] Almost all previous attempts had used indirect signatures of NO production. Direct detection of NO is now possible because of the development in recent years of a number of reliable analytical procedures. These will be reviewed before we proceed to more biological topics, but it is first necessary to describe some of the physical characteristics of NO that have made its quantitation, particularly in biological system, so difficult.

2 Physical Properties of NO

NO is a gas at room temperature, with a low solubility in water (10^{-3} M). From aqueous solution it is rapidly lost to the headspace. In oxygenated water it is oxidized slowly, at a rate which it is critically concentration-dependent, to nitrite (equations 1–3). At this stage no nitrate is formed.[12] However, in oxygenated water and particularly in certain biological media such as blood, nitrite is rapidly oxidized to nitrate.[13]

$$2NO + O_2 \rightarrow 2NO_2 \tag{1}$$

$$NO_2 + NO \rightarrow N_2O_3 \tag{2}$$

$$N_2O_3 + H_2O \rightarrow 2HNO_2 \tag{3}$$

NO is a radical species, but unlike most other radicals, it is not particularly reactive. However, it does react very rapidly with iron to give well authenticated iron-nitrosyls.[14] Reaction with iron is responsible for the activation of guanylate cyclase; it may also be responsible for the cytotoxicity of NO and it has been incorporated into a number of analytical procedures for the detection and quantitation of NO.

NO has high diffusion rates in both aqueous and lipid media, and travels rapidly between cells and across phospholipid membranes.[15] It has a biological half-life of less than a minute[2] and on oxidation forms relatively stable anions.[10,12] This instability *in vivo* reflects reactions with oxygen, superoxide and haemoglobin, and these reactions limit the volume through which NO can travel to exert influence.[16,17]

The low molecular weight of NO may allow rapid movement in aqueous environments. Molecules larger than NO have much lower diffusion coefficients and rates of spread. Models to describe the kinetic and concentration profiles for NO have been developed, based on diffusion models used to describe heat conduction in solids.[17] These models show NO concentrations to be a function of the rate of formation, diffusion coefficient, distance from the source and time. Close to a production source, perhaps within a radius of $10\,\mu m$, concentrations reach a steady state within a few hundred ms of release. This is an order of magnitude faster than the measured biological half-life of NO, so that metabolism of NO *in vivo* should not significantly affect these concentration gradients established by diffusion. The metabolism of NO is likely to become more significant the greater the distance from the source of production. Within a close range, far more NO will be lost to a given site by diffusion than by metabolism. NO is able to influence cellular function over spherical volumes with an influence that is both predictable and constant. Close to a production source, this influence will depend on the diffusion properties of NO itself. The inverse relationship that exists between the half-life and the concentration of NO is likely to become increasingly important to biological functions the farther NO travels from the source of production. The reaction between oxygen and NO is a third-order process (equations 1–3) and will be very slow at low concentrations of NO. Thus the low levels of NO involved in bioregulation permit survival of the molecule in the presence of oxygen.

Unfortunately, some of the same physiochemical characteristics make quantitation of NO difficult. Most studies *in vivo* and *in vitro* have used indirect indices of NO production such as breakdown products of NO metabolism, second messengers, or biological events in effector systems. Recent developments in probe technology and mass spectrometry have allowed the direct detection of NO in some situations. Constitutive NO synthases (cNOS) in vascular endothelium and other tissues produce small quantities of NO for continuous maintenance of vascular tone and cellular communication.[10] Inducible NO synthases (iNOS) are induced by immunological stimuli such as interleukins[18] and lipopolysaccharides, and synthesize relatively large amounts of NO for long periods.[19] In general, therefore, it is much more difficult to study constitutive than inducible NO production.

The complex fate of NO biological systems helps confound investigation and analysis. The amount of NO required for biological action never exceeds the solubility of NO in water, and so the aqueous chemistry of NO is important. More specifically, it is the chemistry of NO in complex biological media, particularly whole blood, that is more important to understand.

3   Measurement of NO in Biological and Chemical Systems

Direct Measurement

Direct measurements of NO are attractive because they offer the highest specificity. The most direct approaches are detection of electric current produced when NO is oxidized, detection of light produced when NO reacts with ozone, and direct GC–MS detection of NO.

By use of the first effect, two probe systems have been developed for detection of NO, and they have been valuable for *in vitro* work. Probes can be used to determine the exact locality of NO and to follow in real time the kinetics of a system. Judicial positioning of more than one probe in a system can be used to assess direction and speed of travel.[15] The specificity of NO probes depends on their ability to exclude other molecules that may give rise to a signal. Shibuki has developed an electrochemical sensor based on a modified oxygen electrode

(Clark electrode) consisting of platinum wire as the anode and silver wire as the cathode.[20] The response time of this sensor is 3–6 seconds, the detection limit $5 \times 10^{-7}\,\mathrm{mol\,l^{-1}}$ and range of linearity 1×10^{-6} to $3 \times 10^{-4}\,\mathrm{mol\,l^{-1}}$. This electrode has been applied to measurement of NO in the central nervous system *in vitro*. Unfortunately, detection limits of some commercial forms of this probe may be no lower than $5 \times 10^{-6}\,\mathrm{mol\,l^{-1}}$; these would be unsuitable for work in many physiological systems. Most versions of this probe are several millimetres in diameter and so will not allow particularly accurate positioning in a cellular system. In complex media, such as plasma, interference and delays in signal equilibration are further problems. Malinski[15] has developed a detection system in which NO is oxidized on a polymeric metalloporphyrin. This porphyrinic semiconductor is covered with a cation exchanger, Nafion, which eliminates anionic interference from nitrite. The high sensitivity, small diameter $(0.2–1.0\,\mu\mathrm{m})$, and fast response time $(10\,\mathrm{ms})$ are useful features for detection of NO in microsystems such as single cells; these are commonly grown in culture for experimental purposes. The detection limits may be as low as $1 \times 10^{-8}\,\mathrm{mol\,l^{-1}}$, and the current concentration relationship is linear between 1×10^{-8} and $3 \times 10^{-3}\,\mathrm{mol\,l^{-1}}$, a wider range than for the Clark electrode. The Malinski microsensor has been used to demonstrate travel of NO from endothelium to smooth muscle and the release of NO by platelets during aggregation. There have also been successful *in vivo* applications of the Malinski probe, including measurement of NO in rat brain during ischaemia. It is a very delicate system and requires considerable technical skill to build and operate. The paucity of applications to have emerged since the enthusiastic introduction of this probe in 1992 testifies to that.

In general, probes do not offer the potential for satisfactory measurement of systemic NO production *in vivo* in humans. They would probably be too delicate to place *in vivo* routinely; they are invasive; and they are predominantly influenced by NO in their immediate vicinity and thus represent local NO production, which is not necessarily representative of systematic constitutive production. They may be of greater value to those investigating the chemistry of NO in aqueous media.

The other direct forms of NO measurement are chemiluminescence and mass spectrometry. These are discussed in a later section in relation to investigations of the respiratory system.

Indirect Measurement

There is a wide range of indirect indices for NO production. These are most commonly used for the investigation of NO in biological systems. They do not have the specificity of direct measurements, but they are more convenient. They include NO metabolites, the second messenger cyclic guanosine monophosphate (cGMP), and effector organ responses such as vessel dilation and platelet function.

The earliest measurement of NO involved bioassays. These are biological systems able to provide units of response in an effector system, such as the degree of vessel dilation or platelet aggregation. The endothelium-derived relaxing factor (EDRF) was first observed using a tension device and a ring of rabbit aorta.[21] It was identified as NO in a bioassay system of vascular strips, by experiments using a cascade of strips of endothelium-denuded artery.[2] Bioassay studies have recently reduced any controversy over the chemical identity of the EDRF.[22] Since many of the biological parameters regulated by NO, such as vessel tone, are also controlled by other molecules, specificity is a potential problem with bioassay. Steps should be taken to assess the specificity of each system, including inhibition of NO production and inhibition of other factors with similar biological actions. Bioassay is a *qualitative* measure and does not provide *quantitative* evaluations of NO.

NO stimulates soluble guanylate cyclase to produce, from guanosine triphosphate, cyclic guanosine monophosphate (cGMP), an intracellular messenger that interacts with a range of receptor proteins to produce its effects. Concentrations of cGMP have therefore been used as indices of NO production.[23] Unfortunately, there are many different forms of guanylate cyclase, and many molecules activate these enzymes, increasing cGMP production. These molecules include all naturetic peptides so far identified (atrial, B type, C type), NO, carbon monoxide, nitrosothiols, and hydroxyl radicals. Cyclic GMP concentrations can also be influenced indirectly by substances that inhibit phosphodiesterase. The specificity of changes in cGMP as an index of NO production may therefore be poor. In contrast with bioassay systems, the low specificity of cGMP is difficult to improve, because it is not possible to inhibit selectively most of the molecules that activate guanylate cyclase.

Analysis of nitrite and nitrate in biological samples has been performed for many reasons. Much of the early interest was related to nitrosamine formation and the potential of these compounds in cancer causation, but recent interest in NO as a mediator cell function has greatly increased interest in measurement of these anions. As a measure of NO, this measurement works well in aqueous solution in the presence of oxygen, where conversion of NO into nitrite is quantitative and other sources of nitrite apart from NO can easily be controlled.[12] The sole formation of nitrite is surprising, as hydrolysis of NO_2, the oxidation product of NO, gives a mixture of nitrite and nitrate. It seems that, in an aqueous solution, the reaction of NO_2 with NO to give N_2O_3, the anhydride of nitrous acid, is faster than is the hydrolysis of NO_2. However, in biological fluids it is normal to obtain a mixture of nitrite and nitrate from NO. Many factors, including superoxide, haemoglobin and hydroxyl radicals, may affect *in vivo* NO metabolism. Recent studies suggest that NO reacts rapidly with superoxide anion to form peroxynitrite, $ONOO^-$, which is protonated at physiological pH.[24] (This important process is discussed in detail later.) Peroxynitrite readily isomerises to nitrate (equation 4). NO also will react with

$$ONOO^- \rightarrow NO_3^- \qquad (4)$$

oxyhaemoglobin to produce methaemoglobin and nitrate. Thus, there are several potential routes for formation of nitrate from NO. An additional aspect to the use of nitrite and nitrate as indices of NO production is that sources of nitrite and nitrate other than NO are difficult to control *in vivo*. Therefore, it is important to measure both anions in most circumstances; ideally, they should be measured simultaneously. There are specific problems with using nitrite and nitrate concentrations in whole blood to assess NO formation *in vivo*. Nitrite is not particularly stable in whole blood, and may be oxidized to nitrate.[13] Thus the time delay before centrifugation of the whole blood sample influences the plasma nitrite levels subsequently measured. Moreover, nitrite can probably be formed in plasma after sampling, by reaction of less stable NO metabolites such as *S*-nitrosothiols.[25] Duration of sample storage prior to analysis should be standardized to achieve maximum reproducibility. Nitrate is stable in whole blood and plasma, and because it is present there in much higher concentrations than is nitrite, small nitrate contributions from breakdown of less stable NO metabolites are probably not as important as are those to nitrite. However, the half-life of nitrate *in vivo* is believed to be several hours, and it may therefore be chronologically insensitive to small, acute fluctuations in NO production, particularly reductions in the NO production rate. The ubiquitous nature of these anions calls for specific dietary preparation of subjects before study, and defined methods of venesection, storage, and sample handling. Any assay for the contribution of nitrite and nitrate that is to be applied to biological work must be assessed by its ability to detect nanomolar, or at least micromolar, levels in the presence of millimolar chloride and sulfate. A commonly used method[26] for the assay of biological nitrate is gas chromatography–mass spectrometry (GC–MS), a method based on production of nitroaromatics from benzene or related compounds in strong acid. Although this approach has been extensively used, it unfortunately leads to aromatic production from a wide range of nitrosothiols present in plasma, not simply from nitrate. Of the wide range of alternative methods, by far the most valuable for biological work is capillary electrophoresis, which permits rapid and virtually contamination-free simultaneous analysis of nitrite and nitrate in plasma at physiological and subphysiological levels.[27]

Part of the controversy over the identity of the endothelium-derived relaxant factor relates to *S*-nitrosothiols (such as *S*-nitrosocysteine), which decompose in solution to give NO that is converted into nitrite in the presence of oxygen.[28]

A variety of *S*-nitrosothiols is believed to be present in human plasma,[29] and endogenous *S*-nitrosothiols have been proposed for a variety of biological functions. The mechanisms by which these compounds are formed and their biological roles have not yet been fully described. It is possible that nitrosothiols are generated *in vivo* from the active species produced from NO, such as peroxynitrite. Whether these nitrosothiols function as a means to remove or donate NO is unclear. They may constitute an elimination mechanism, sequestering and inactivating NO; they may form a reservoir that can be used to provide NO. For these reasons it is not possible to predict whether any of the nitrosothiols are likely to provide reliable indicators of NO production. Low-molecular-weight thiols and *S*-nitrosated derivatives are difficult to analyse, and this problem has also contributed to the slow pace of their assessment in biological systems.

The complex formed between NO and haem iron (HbNO) has well-defined electron paramagnetic resonance (EPR) spectroscopic properties.[30] The specificity of the HbNO assay is good because the characteristic three-line hyperfine pattern produced is not seen when haemoglobin binds nitrite and other nitrogen-containing compounds. A number of animal models have applied HbNO as an index of NO production. Increases in HbNO have been shown using EPR after inhalation of NO and in mouse and rat models of sepsis. However, it is not clear in which circumstances HbNO may provide a good index of NO production. Most indices are measured in moles per unit volume, and changes reflect not only production but also metabolism, an important consideration when assessing the meaning of changes in an index like HbNO, because its metabolic fate is not known. Indeed, this remains one of the fundamental questions about the chemistry and biology of NO. It is certainly important that NO haemoglobin does not accumulate in the circulation, because if it did the oxygen-carrying capacity of the blood would be rapidly compromised. When human subjects are exposed to carbon monoxide, the formation of a complex between carbon monoxide and haemoglobin can give rise to respiratory failure and, with sufficient exposure, to death. NO has very high affinity for haemoglobin, greater than the affinities of oxygen or carbon monoxide, and one would therefore expect HbNO to accumulate in the body. Yet, despite constant production of NO in the body that does not happen with the NO–haemoglobin complex. Clearly, there are mechanisms to prevent this potentially fatal complex from accumulating.

It has been suggested that redox forms of NO, the nitrosonium ion NO^+ and the nitroxide ion NO^-, may have physiological roles that are conflated with those of NO. If this were so, there could be important ramifications for the variety of indices of NO production currently in use among biologists. The matter will be considered in more detail later in this review. The relevant chemistry of these species has been reviewed.[31]

4 Nitric Oxide and the Respiratory System

Work with NO and the respiratory system provides the only direct evidence of NO production *in vivo* in humans,[11] and analytical techniques devised for the detection of NO as an atmospheric pollutant have been applied. Reaction between NO and ozone leads to the generation of light (equations 5 and 6).

$$NO + O_3 \rightarrow NO_2^* + O_2 \tag{5}$$
$$NO_2^* \rightarrow NO_2 + light \tag{6}$$

A variety of chemiluminescence analysers is now available for quantitation of NO, and such equipment can be adapted readily for routine measurement of endogenous NO excreted in expired air.[11,32] The technique is highly sensitive (detection limit 1 ppb) and linear over a wide range; 1 to 1000 ppb can easily be covered without recalibration of equipment. Analyses are rapid and highly reproducible. Further work is still required to establish the exact origins of NO contributing to exhalation profiles recorded with this technology. Both lower and upper respiratory tracts release NO into expirate.[33,34]

Mass spectrometry has been used to confirm the chemiluminescence analyses of NO in breath. This can be demonstrated indirectly after activation by oxidation, using the nitrosation of thioproline and analysis of the nitrosothioproline derivative by GC–MS.[35] GC–MS confirmation of the presence of NO in the breath in humans[11] demands differentiation between $^{15}N_2$ (m/z 30.00022, present in parts per thousand) and NO (m/z 29.99799, present in parts per billion). Direct GC–MS analysis of NO *in vivo* is complex, temperamental, and not practical for routine use. Whether or not useful representation of systematic NO production can be obtained from breath analyses has still to be determined. At present, it seems likely that the measurement of NO in breath will continue to be a useful tool for investigation of systematic NO generation. The response time of chemiluminescence equipment is rapid (< 0.5 s), so exhaled NO levels can be determined in a single breath.[33] This method has been used to detect changes in NO production that occur during exercise. There is good reason to believe that systematic NO production rises during exercise. The physiology cardiovascular changes that occur in exercise are profound, and increases in blood flow and sheer stress, as well as reduction in pO_2, are known to elevate NO release from the endothelium.[36,37]

The measurement of NO in breath by chemiluminescence has been used to access exhaled NO in asthma and primary pulmonary hypertension (PPH). The technique is not suitable for assessment of patients with poor control over their respiratory cycle, since a slow (4 rpm), regular breathing pattern with prolonged exhalation time (> 5 seconds) is necessary to create the NO plateau required for standardized measurements. Breathing with normal to low tidal volumes (> 1 litre) and normal to high frequency (> 8 rpm) does not give rise to a stable end expiratory concentration of NO. Therefore, assessment of NO in the expirate of subjects with chronic dyspnoea or neutromuscular disorders would be difficult by these means. Asthmatics exhale greater than normal amounts of NO and express inducible NO synthase in lung tissue. The implications of these findings are not clear. It is likely that raised values reflect the activation of inducible NOS in the airways as part of a generalized inflammatory response.[38]

NO is able to cause vascular and bronchial smooth muscle relaxation. The principal mode of action is stimulation of soluble guanylate cyclase with consequent rises in intracellular cyclic GMP. Inhalation of NO gives selective reduction in pulmonary vascular resistance in pulmonary hypertension[39] and in adult respiratory distress syndrome (ARDS).[40] ARDS is a severe and often fatal condition associated with pulmonary oedema and reduced right ventricular function. It is often treated with vasodilators to reduce pulmonary artery pressure, but unfortunately general systemic vasodilation also occurs, reducing systemic arterial pressure. An additional disadvantage of this approach is that systematically infused vasodilators dilate the entire pulmonary vascular bed, augmenting blood flow to non-ventilated or poorly ventilated lung regions, increasing pulmonary venous mixing and reducing arterial pO_2. However, prolonged administration of NO followed by withdrawal gives sustained improved lung function, suggesting improved blood flow and oxygen exchange, and leads to recovery of damaged tissue. This is an exciting development in the treatment of lung disease. Not only does inhaled NO give superior pulmonary vasodilation as compared with systematically administered vasodilators, but NO, as a gas, reaches only parts of the lung still functioning, so ventilation-perfusion mismatching does not occur, and the risk of reduced pO_2 is avoided. The full potential of inhaled NO for treatment of pulmonary hypertension and ARDS will be realized only after full clinical trials. The risks involved relating to NO_2 formation and methaemoglobinaemia appear real.

5   Nitric Oxide and the Cardiovascular System

Our perception of the cardiovascular system has changed over recent decades, from one of a rigid system of conducting pipes predominantly acted upon by vasoconstrictors, to one of a highly flexible tree under constant regulation by vasoconstrictor and vasodilator forces. We now know that in the testing state the vascular system is under a profound, constant vasodilator influence.

The enzymes responsible for NO production can be inhibited competitively by analogues of L-arginine such as N-monomethyl-L-arginine (LNMMA).[10] When such an analogue is infused into the forearms of normal subjects, profound vasoconstriction occurs, resulting in as much as a 50% fall in basal blood flow. This work[41] demonstrates the continuous release of NO in the forearm arterial bed and its major role in determining basal blood flow, as well as modifications in blood flow patterns with appropriate stimuli. The vasoconstricting effect in the human forearm achieved with LNMMA cannot be reproduced with its enantiomer but can be reversed with L-arginine, confirming enantiomeric specificity. Theoretical calculations suggest that without reflex constrictor changes, inhibition of NO production in the vasculature would result in a rise in total peripheral resistance sufficient to double the mean arterial blood pressure. It therefore appears that production of NO by the vascular endothelium is of great functional importance. The magnitude of the response seen after inhibition of NO synthesis is particularly striking when compared with the minor changes of blood flow that can be achieved by manipulation of many other vasoactive hormones in humans.

There are a number of physiological stimuli for NO release by the endothelium that help to regulate blood flow under normal physiological conditions. These include the flow and amplitude of wave forms created by the passage of blood through the vessel and the partial pressure of oxygen.[42] Profound changes in blood flow occur during exercise, with blood directed towards active muscle groups and away from large viscera such as the gastrointestinal tract. NO is likely to take part in such profound redirection of blood. There is also some recent work[43] involving chronic exercise of animals that suggests that increased expression of NOS occurs in vascular endothelium as a response to the repeated stimulus of exercise.

Septic shock is characterized by progressive reduction in systemic blood pressure and resistance to exogenously administered vasoconstrictors, poor tissue perfusion, and vascular leakage. The basis of current therapy is the maintenance of tissue perfusion pressure with plasma expanders that raise colloidal osmotic pressure in the intravascular space, and with catecholamines that cause peripheral vasoconstriction and increased cardiac contractility. The mortality of this condition remains high. It is now believed that excessive NO production occurs in septic shock and contributes to pathological changes. The potential therapeutic benefit of inhibitors of NOS has recently been demonstrated in a small number of patients with septic shock.[44] Larger trials will be required to establish the benefits and the correct administration protocol for such patients. A great deal of research effort is currently under way, particularly in the pharmaceutical industry, to pursue specific inhibitors of inducible NOS. Such a therapeutic approach would be exciting, allowing constitutive physiological NO production to continue to fulfil its normal function, while preventing excessive release of NO from pathological expressed iNOS.

Pre-menopausal women suffer less cardiovascular disease than men do. This protection disappears, however, after the menopause, and hormone replacement therapy in post-menopausal women reduces cardiovascular mortality. Animal work suggests that oestrogens dilate blood vessels by an endothelium-dependent mechanism, but as yet there is no direct evidence to show that NO generation is different between men and women. This is an area of considerable interest, particularly as women have greater longevity than men and suffer less ischaemic heart disease. Abnormal NO production may occur in hypercholesterolaemia and may be related to subsequent development of atherosclerosis. There is now a body of experimental data to suggest that an abnormality in NO production or function may be causal, or at least an amplifying factor, in both hypercholesterolaemia and atherosclerosis.

During the last century a role for nitroglycerin in the treatment of acute attacks of cardiac angina was suggested. It has been the mainstay of anti-anginal therapy ever since; yet only in recent years has the chemical basis of this therapy become clear. Nitroglycerin is an NO donor, metabolized to NO in the vasculature. It, and other organic nitrates such as isosorbide dintrate and isosorbide mononitrate, may be of particular benefit in the presence of coronary atherosclerosis, hypercholesterolaemia, and other conditions associated with endothelial dysfunction. There is considerable evidence to suggest that the site of metabolic conversion into NO is the vascular smooth muscle.[45]

There have been suggestions that nitroglycerin can substitute for defective NO production in atherosclerosis. There are important differences, however, between nitroglycerin and endogenous, endothelium-derived NO. Whereas endogenous NO and a number of other NO-donor drugs dilate all classes of coronary microvessels, nitroglycerin has only minor effects on coronary microvessels less than 100 μm in diameter. Studies of the coronary microcirculation both *in vitro* and *in vivo* suggest that the smaller coronary vessels are incapable of converting nitroglycerin into its vasodilator metabolite.[46] Other NO-donor compounds, such as sodium nitroprusside and nitrosothiols, are not dependent on the biotransformation needed for release of NO from nitroglycerin and are potent vasodilators of all microvessels. The favourable anti-ischaemic effect of nitrates is based on this unique pattern of vascular relaxation. Nitrovasodilators reduce cardiac preload and wall tension and thus myocardial oxygen consumption. They increase precolateral perfusion pressure, augmenting oxygen delivery to ischaemic sections. The differential vasomotor responses result in a unique pattern of therapeutic uses and relevant drug action; dilation of the venous bed giving preload reduction, dilation of large coronary arteries improving conductivity and blood supply, and the absence of effects on vessels that would unfavourably reduce perfusion pressure. These differential effects are the most challenging feature of organic nitrates, against which any new NO-donor drug must compete. The generalized vasodilation caused by many NO-donor drugs is not wanted in most clinical situations. New NO-donor drugs for anti-anginal therapy must therefore be evaluated for their effects on the coronary microcirculation. It is quite possible that chemists will be able to develop compounds that mimic the pattern of action nitroglycerin in the coronary microcirculation, and these compounds may be useful in the treatment of regional myocardial ischaemia. In contrast, NO-donor drugs that produce uniform dilation of all sizes of coronary microvessels will not be effective and may well cause coronary steal, worsening regional ischaemia.

One of the major disadvantages of the organic nitrates as therapeutic agents is the development of tolerance.[46,47] The metabolic step leading to nitrate tolerance is believed to precede the final formation of NO *in vivo*. Delivery of NO spontaneously from NO-donor drugs and continuous stimulation of endogenous NO production by increased blood flow both lead to the desensitization that is witnessed during continuous nitrate administration. Impaired nitrate metabolism within the vasculature is an important aspect of organic nitrate tolerance. In addition, it is likely that another feature of impaired metabolism is neurohormonal activation. The combination of these two aspects would reduce vasodilation and induce sodium ion and water retention. These reflex neurohormonal changes reduce the dynamic effect of nitrate-induced vasodilation. Although preliminary studies[48] suggest that diuretic therapy or enzyme inhibitors may reduce tolerance, further clinical studies are required to consolidate the information. Nitrate tolerance is an important clinical problem, and the effectiveness of nitrate therapy is markedly reduced during treatment regimes maintained throughout a 24 hour period. Strategies for avoiding tolerance are usually based on a nitrate-free interval. Tolerance to nitroglycerin and isosorbide dinitrate can be compensated for by administration of a different NO-donor such as SIN-1, which spontaneously releases NO and does not show significant cross-tolerance with organic nitrates. However, when applied at high concentrations for several days, even SIN-1 induces some tolerance.

Although the main role of NO-donor drugs has been in treatment of cardiac angina, the pain caused by myocardial ischaemia, there are other potential benefits to be had from NO-donor drugs in treatment of myocardial ischaemia and infarction. These agents are also fundamental to the treatment of acute left ventricular failure, where they reduce preload on the heart and dilate the venous system, allowing improved cardiac output. They also have potential roles in the treatment of hypercholesterolaemia and atherosclerosis. It has been claimed that they reduce uterine contractions in premature labour[49] but confirmation is lacking.

There are alternatives to organic nitrates as NO-donor drugs, and much interest in developing new NO-donor drugs exists. However, the organic nitrates present formidable therapeutic competition because of their specific regional metabolism and ideal 'design' for the treatment of cardiac ischaemia. The tolerance developed to organic nitrate therapy may be an area amenable to new product development, but the problem should not be overestimated. In most patients it can be adequately managed with a rest period from the therapy at night.

6   Nitric Oxide and the Nervous System

It has been known for over two decades that synaptic excitation in the central nervous system (CNS) is accompanied by an increase in cGMP. This phenomenon is now known to be caused by NO, and a major trigger for its release in the nervous system is the activation of glutamate receptors.[16] The availability of agents to supply exogenous NO and to inhibit NOS has proved essential for investigation of these effects. The most direct demonstrations of NO production in the nervous system to date have come from use of the Shibuki probe.[50] The clearest role for the NO has been established in the peripheral autonomic nervous system, where NO is released from non-adrenergic, non-cholinergic (NANC) nerves and mediates relaxation of smooth muscle in many tissues.[51] For the CNS, most interest has revolved around the three main functions of synaptic plasticity, regulation of blood flow, and neurodegeneration.

Peripheral Nervous System

Inhibitory NANC nerves are recognized as important components of autonomic innervation to many organs. The chemical identity of the transmitter has been much debated, but convincing demonstration that NO fills this function has now emerged, most recently by comparison between diffusion properties of the NANC nerve transmitter, and those of NO itself. These diffusion properties are highly characteristic of NO and could not be achieved by nitrosothiols or other contenders for the NANC nerve transmitter.[17] Superfusion bioassay has shown release of a vasorelaxant factor with properties similar to those of NO, upon stimulation of NANC nerves in canine ileocolonic function. In the gastrointestinal tract, NO seems to mediate many forms of relaxation, including adaptive dilation of the stomach following ingestion of food.[52] Histochemical studies of biopsies from infants with hypertrophic plyoric stenosis and adults with achalasia suggest that these conditions are caused by a lack of NOS in pyloric and gastro-oesophageal tissue. In the intestine, muscle relaxation involved in peristalsis is also mediated by NO.[53] Therefore, in the gastrointestinal tract, as in the cardiovascular system, there seems to be a constant NO-dependent dilator tone crucial for function of these organs.

NO is responsible for relaxation of the blood vessels and smooth muscle of the male corpus cavernosum and thus development of penile erection. Immunohistochemical evidence of NO-containing nerves has been found in penile tissue.[54] Other work has indicated dysfunction of the NO system in disorders associated with male impotence.[55] During erection, active filling of the sinusoids compresses the peripheral venules of the corpora against the rigid tunica albuginea, causing outflow obstruction, but this process is not mediated simply by passive engorgement with blood. Flaccidity is enhanced by contraction of corporal tissue after activation of adrenergic receptors by noradreline released from sympathetic nerve terminals. Functional studies with isolated human and animal corpus cavernosum tissue have shown that relaxation induced by electrical stimulation of autonomic nerves is abolished by blockade of NOS.[56] Immunohistochemical staining of both rat and canine tissue has shown that NOS is widely distributed throughout the urogenital tract, high concentrations being found in major pelvic ganglia, the membranous urethra, and bladder neck, as well as in the penis itself.[54,57] The high functional NOS activity found in the membranous urethra and its involvement in relaxation of the bladder neck further suggests that NO may be important in the regulation of micturition and urine continence.

Involvement of the NO-arginine pathway in penile erection may explain the aphrodisiac affects of amyl nitrite. In impotent diabetic men, there is an impairment of both neurogenic and endothelium-mediated relaxation in penile corporal smooth muscle, whereas responses to exogenous donors of NO are preserved. The penile flaccidity found in diabetes may be a direct consequence of impaired NO synthesis; from a clinical perspective, this provides a rationale for the treatment of certain types of impotence by intracavernosal administration of vasodilators. Conversely, it is possible that inhibition of NOS will be beneficial in the treatment of priapism. NO also contributes to NANC-induced vasodilation and relaxation of guinea pig and human tracheal muscle.[58] Clearly there is a widespread system of nerves throughout the body using NO as a transmitter, and it is proving to be as important as those in adrenergic, cholinergic and peptidergic nerves.

Synaptic Plasticity

Synaptic plasticity is a process by which neuronal connections may be re-enforced or altered. It is involved in cerebral functions such as memory and in complex automated motor functions, such as playing a piano concerto fluently. Part of this phenomenon involves retrograde messengers that return from a stimulated neurone to relay information to the preceding neurone. NO is an attractive candidate for a retrograde messenger mediating use-dependent changes in synaptic transmission by acting at pre-synaptic sites following post-synaptic production.[59] The most studied example of synaptic plasticity is long-term potentiation in the hypocampus, a region of the brain involved in memory and learning, but the role of NO in this phenomenon remains controversial and uncertain.

Another example of synaptic plasticity in hyperalgesia, a state of enhanced sensitivity to painful stimuli. The mechanism of hyperalgesia shares several features with hypocampal long-term potentiation, particularly its endurance, its dependence on afferent fibre activity, and its blockade by (NMDA) antagonists. A range of evidence suggests a role for NO in the plastic changes giving rise to the hyperalgesia, but further experimental evidence is required to understand the basic mechanisms involved.[16,60,61]

Tolerance to morphine is a further form of synaptic plasticity in which NO may be involved. The mechanisms of morphine tolerance are poorly understood, but glutamate acting at NMDA receptors is probably involved, because development of tolerance can be inhibited by NMDA antagonists. It has recently been shown[62] that NOS inhibitors given with morphine completely prevent tolerance without affecting the actions of morphine itself, suggesting a selective effect on the tolerance mechanism.

Cerebral Circulation

The possibility that neuronally-derived NO may be important in the regulation of cerebral circulation has received much attention. It is believed to act within cardiovascular centres in the CNS to regulate sympathetic outflow, vascular tone, and arterial pressure.[63,64] NOS-containing fibres originating in one of the cranial autonomic ganglia innovate major cerebral muscles and provide neurogenic dilator control over the cerebral circulation.[65] NO generated by neurones within the brain parenchyma may also be a factor that links the neuronal activity to local increases in cerebral blood flow. These mechanisms may all be significant, and participation of NO in the maintenance of resting cerebrovascular tone, as well as dilator responses to stimuli appear to be fundamental mechanisms.[66] One of the most puzzling and damaging consequences of subarachnoid haemorrhage is vapospasm. Subarachnoid haemorrhage, or indeed any form of intracranial haemorrhage, releases red blood cells from their intravascular compartment. It is possible that this could lead to sequestering of NO by free haemoglobin, and vasoconstriction in the region of the bleed. That would cause the vasospasm and the consequent cerebral ischaemia responsible for much of the damage that can occur in the recovery phase following intracranial haemorrhage.

Neurodegeneration

The possibility that NO can contribute to neurodegeneration in the CNS has been raised, because it is a potentially toxic molecule and is able to inactivate key enzymes either directly or by peroxynitrite formation.[24] In cultures, activated microglias, macrophage-like cells in the CNS, can produce enough NO to kill cocultured cerebral neurones,[67] and it is therefore possible that NO production *in vivo* could lead to neuronal damage. In *in vivo* models of focal cerebral ischaemia, NOS inhibitors have been found to give marked protection, rescuing up to 70% of cortical neurones in some studies.[68] However, these inhibitors can also increase damage.[69] This is an exciting area where the chemical synthesis of selected inhibitors of neuronal NOS is greatly needed, to avoid the potentially detrimental action of blocking vital NO-dependent mechanisms, such as cerebral blood vessel dilation and platelet de-aggregation.

In view of the apparently conflicting roles of NO in the brain (it can be both neurotoxic and neuroprotective), it has been proposed[70] that the redox forms of NO (*i.e.* NO^+ and NO^-) play parts in its activity. Although that presents an attractive solution to the paradox of NO behaviour, careful examination of the chemistry involved reveals difficulties. NO^+ cannot exist in aqueous solution, as there is very rapid hydrolysis to nitrous acid (equation 7). However, nitrosothiols can transfer NO^+ to other nucleophilic sites; within a cell this probably means to another thiol. Therefore, NO^+ could be said to interact with physiologically significant thiols. NO itself does not react with thiols to give nitrosothiols, but there is reaction in the presence of an oxidizing agent. Thus, it is possible to see how nitrosothiols are formed *in vivo*, but the physiological importance of trans-nitrosation to another thiol has not been delineated. The nitroxide ion NO^- is formed by ionization of HNO (equation 8), but there is competing dimerization of HNO to give nitrous oxide (equation 9). So far, there is not direct evidence for the involvement of NO^d in physiological processes.

$$NO^+ + H_2O \rightarrow HNO_2 + H^+ \tag{7}$$

$$HNO \rightleftharpoons H^+ + NO^- \tag{8}$$

$$HNO + HNO \rightarrow N_2O + H_2O \tag{9}$$

7   Nitric Oxide and the Immune System

The immune system provides a range of defence mechanisms that protect against foreign proteins and invasion of the intracellular environment by micro-organisms. The system produces a number of reactive oxygen intermediates (ROIs) in response to inflammatory stimuli.[71] As has already been mentioned in the Introduction, it now appears that NO is also involved in the physiology of the immune system. The first indication of this came when it was noted[72] that high levels of nitrate were found in subjects suffering from gastroenteritis. Treatment of rats with *E. coli* lipopolysaccharide also leads to high nitrate levels. Demonstrations then followed that showed that cultured macrophages, one of the main cell groups involved in the immune protective system also produce nitrate.[4-6] This process is dependent upon the presence of arginine.[73] These observations led to the conclusion that the arginine-NO pathway occurs in activated macrophages. The ability of macrophages to kill abnormal cells, such as tumor cells, was then shown to be dependent upon arginine.[74] NO must therefore be a natural cytotoxic agent for the defence of the body against foreign agents and against abnormal cells within the body.

The question now engaging the attention of scientists in a number of disciplines is the nature of the cytotoxic action of NO. As NO is a radical, it was initially suggested that this might be sufficient, but unlike most radicals, NO is not particularly reactive. When tested against *Clostridium*, an aqueous solution of NO is not a particularly powerful antibacterial agent.[75] As macrophage activity also produces ROIs, it has been proposed that NO produced by macrophages reacts with superoxide to give the peroxynitrite ion (equation 10).[76] This ion is a relatively strong base[77] and is protonated at physiological pH (equation 11). Peroxynitrous acid then decomposes to give nitrogen dioxide and hydroxyl radicals, which are certainly highly reactive (equation 12).[78] However, not all the experimental evidence supports this view. When tested against *Leishmania major*, S-nitro-N-acetylpencillamine (SNAP) was found to be an effective toxic agent, the effect of which was enhanced by addition of superoxide dismutase (SOD), which removes any superoxide present. In contrast, peroxynitrite was not particularly toxic.[79] It may be that the reaction of peroxynitrite with cellular targets, rather than the formation of hydroxyl radicals, is responsible for its biological activity, and this would depend upon the environment in which it finds itself.[80] A similar conclusion was reached in a study of the effect of peroxynitrite on the aggregation of washed platelets.[81]

$$NO + O_2^- \rightarrow ONOO^- \tag{10}$$

$$ONOO^- + H^+ \rightarrow ONOOH \tag{11}$$

$$ONOOH \rightarrow NO_2 + HO\cdot \tag{12}$$

Two studies of the antibacterial action of NO-donor drugs suggest that, under the circumstances of these studies, it is neither NO nor hydroxyl radicals that is responsible for their toxic action. Sodium nitroprusside (SNP) contains NO as a ligand where it is formally present as NO^+; SNP is an effective toxic agent against *Clostridium spirogenes*.[75] NO is also present in Roussin's black salt (RBS) as NO^+, and RBS is the most effective of a range of NO-donor drugs against *E. coli*.[82] The exact significance of this finding is at present unclear, but it may be that the crucial element in the toxic effect of SNP and RBS is transfer of NO^+ to a physiologically significant thiol. Both SNP[83] and RBS[84] are hypotensive agents, and the former has been used in hypotensive surgery. In this role they must undergo chemical[85] or enzymatic[86] reactions to release NO. The key may be the formation, from SNP or RBS, of nitrosothiols that can both transfer NO^+ in a heterolytic process and give NO in a homolytic one.

In view of the complexity of most physiological processes it is refreshing to note that a simple diatomic *inorganic* molecule has such a significant role in so much physiological chemistry. Of course, it is not without precedent; dioxygen is also a simple, diatomic, *inorganic* molecule.

Acknowledgements

Our grateful thanks go to many people who have assisted us in exploring this fascinating area on the borders of chemistry and physiology, but principally we thank Professor Salvador Moncada and his collaborators at The Wellcome Research Laboratories, England, who have been responsible for so much of what has been described.

References

1 S.A. Waldman and F. Murad, *Pharmacol. Rev.*, 1987, **39**, 163.
2 R.M.J. Palmer, A.G. Ferrige and S. Moncada, *Nature*, 1987, **327**, 524.
3 L.J. Ignarro, G.M. Buga, K.S. Wood, R.E. Byrns and G. Chaudhuri, *Proc. Natl. Acad. Sci., USA*, 1987, **84**, 9265.
4 M.A. Marletta, P.S. Yoon, R. Iyengar, C.D. Leaf and J.S. Wishnok, *Biochemistry*, 1988, **27**, 8706.
5 J.B. Hibbs, R.R. Taintor, Z. Vavrin and E.M. Rachlin, *Biochem. Biophys. Res. Commun.*, 1988, **157**, 87.
6 D.J. Stuehr, S.S. Gross, I. Sakuma, R. Levi and C.F. Nathan, *J. Exp. Med.*, 1989, **169**, 1011.
7 R.G. Knowles, M. Palacios, R.M.J. Palmer and S. Moncada, *Proc. Natl. Acad. Sci. USA*, 1989, **86**, 5159.
8 J. Garthwaite, *Tends Neurosci.*, 1991, **14**, 60.
9 A.M. Leone, R.M.J. Palmer, R.G. Knowles, P.L. Francis, D.S. Ashton and S. Moncada, *J. Biol. Chem.*, 1991, **266**, 23790.
10 S. Moncada, R.M.J. Palmer and E.A. Higgs, *Pharmacol. Rev.*, 1991, **43**, 109.
11 A.M. Leone, L.E. Gustafsson, P.L. Francis, M.G. Persson, N.P. Wiklund and S. Moncada, *Biochem. Biophys. Res. Commun.*, 1994, **201**, 883.
12 D.A. Wink, J.F. Darbyshire, R.W. Nims, J.E. Saavedra and P.C. Ford, *Chem. Res. Toxicol.*, 1993, **6**, 23.
13 C. Spagnuolo, P. Rinelli, M. Coletta, E. Chiacone and F. Ascoli, *Biochim. Biophys. Acta*, 1987, **911**, 59.
14 L.N. Kubrina, W.S. Caldwell, P.I. Mordvintcev, I.V. Malenkova and A.F. Vanin, *Biochim. Biophys. Acta*, 1992, **1099**, 233.
15 T. Malinski and Z. Taha, *Nature*, 1992, **358**, 676.
16 J. Garthwaite, *Neurosciences*, 1993, **5**, 171.
17 J. Wood and J. Garthwaite, *Neuropharmacology*, 1994, **33**, 1235.
18 R. Busse and A. Mülsch, *FEBS Lett.*, 1990, **275**, 87.
19 M.W. Radomski, R.M.J. Palmer and S. Moncada, *Proc. Natl. Acad. Sci. USA*, 1990, **87**, 10043.
20 K. Shibuki, *Neurosci. Res.*, 1990, **9**, 69.
21 R.F. Furchgott and J.V. Zawadzki, *Nature*, 1980, **288**, 373.
22 M. Feelisch, M. te Poel, R. Zamora, A. Deussen and S. Moncada, *Nature*, 1994, **368**, 62.
23 K. Hishikawa, T. Nakaki, M. Tsuda, H. Esumi, H. Ohshima, H. Suzuki, T. Saruta and R. Kato, *Jpn. Heart J.*, 1992, **33**, 41.
24 J.S. Beckman, T.W. Beckman, J. Chen, P.A. Marshall and B.A. Freeman, *Proc. Natl. Acad. Sci. USA*, 1990, **87**, 1620.
25 J.S. Stamler, D.I. Simon, J.A. Osborne, M.E. Mullins, O. Jaraki, T. Michel, D.J. Singel

and J. Loscalzo, *Proc. Natl. Acad. Sci. USA*, 1992, **89**, 444.

26 J.W. Tesch, W.R. Rehg and R.E. Sievers, *J. Chromatogr.*, 1976, **126**, 743.

27 A.M. Leone, P.L. Francis, P. Rhodes and S. Moncada, *Biochem. Biophys. Res. Commun.*, 1994, **200**, 951.

28 S.C. Askew, D.J. Barnett, J. McAninly and D.L.H. Williams, *J. Chem. Soc., Perkin Trans.* 2, 1995, 741.

29 J.S. Stamler and J. Loscalzo, *Anal. Chem.*, 1992, **64**, 779.

30 M. Hoshino, K. Ozawa, H. Seki and P.C. Ford, *J. Am. Chem. Soc.*, 1993, **115**, 9568.

31 A.R. Butler, F.W. Flitney and D.L.H. Williams, *Trends in Pharmacol. Sci.*, 1995, **16**, 18.

32 T.M. Griffith, D.H. Edwards, M.J. Lewis, A.C. Newby and A.H. Henderson, *Nature*, 1984, **308**, 645.

33 M.G. Persson, P. Wiklund and L.E. Gustafsson, *Am. Rev. Respir. Dis.*, 1993, **148**, 1210.

34 M.G. Persson, O. Zetterström, V. Agrenius, E. Ihre and L.E. Gustaffson, *Lancet*, 1994, **343**, 146.

35 L.E. Gustaffson, A.M. Leone, M.G. Persson, N.P. Wiklund and S. Moncada, *Biochem. Biophys. Res. Commun.*, 1991, **181**, 852.

36 G.M. Rubanyi, J.C. Romero and P.M. Vanhoutte, *Am. J. Physiol.*, 1986, **250**, H1145.

37 I.R. Hutcheson and T.M. Griffith, *Am. J. Physiol.*, 1991, **261**, H257.

38 Q. Hamid, D.R. Springall, V. Riveros-Moreno, P. Chanez, P. Howarth, A. Redington, J. Bousquet, P. Godard, S. Holgate and J.M. Polak, *Lancet*, 1993, **342**, 1510.

39 J. Pepke-Zaba, T.W. Higenbottam, A.T. Dinh-Xuan, D. Stone and J. Wallwork, *Lancet*, 1991, **338**, 1173.

40 C. Frostell, M.D. Fratacci, J.C. Wain, R. Jones and W.M. Zapol, *Circulation*, 1991, **83**, 2038.

41 P. Vallance, J. Collier and S. Moncada, *Lancet*, 1989, **28**, 997.

42 V. Phol, R. Busse and E. Bassenge, *Hypertension*, 1986, **8**, 37.

43 W.C. Seea, K. Pritchard, N. Seyedi, J. Wang and T.H. Hintze, *Circulation Res.*, 1994, **74**, 349.

44 A. Petros, D. Bennett and P. Vallance, *Lancet*, 1991, **338**, 1557.

45 J. Abrams, *Am. J. Med.*, 1991, **91**, 106.

46 E. Bassenge and J. Zanzinger, *Am. J. Cardiol.*, 1992, **70**, 23B.

47 P. Needleman and E.M. Johnson, *J. Pharmacol. Exptl. Ther.*, 1973, **184**, 709.

48 J.D. Horowitz, *Am. J. Cardiol.*, 1992, 64B.

49 C. Lees, S. Campbell, E. Jauniax, R. Brown, B. Ramsay, D. Gibb, S. Moncada and J.F. Martin, *Lancet*, 1994, **343**, 1325.

50 K. Shibuki, *Neurosci. Res.*, 1990, **9**, 69.

51 H. Bult, G.E. Boeckxstaens, P.A. Pelckmans, F.H. Jordaens, Y.M. Van Maercke and A.G. Herman, *Nature*, 1990, **345**, 346.

52 R.A. Lefebvre, E. Baert and J.A. Barbier, in *The Biology of Nitric Oxide*, ed. Moncada, Marletta, Hibbs and Higgs, 1992, Vol. 1, p. 293.

53 S.M. Ward, H.H. Dalzeil, K.D. Thronbury, D.P. Westfall and K.M. Sanders, in *The Biology of Nitric Oxide*, ed. Moncada, Marletta, Hibbs and Higgs, 1992, Vol. 1, p. 295.

54 A.L. Burnett, C.J. Lowenstein, D.S. Bredt, T.S.K. Chang and S.H. Snyder, *Science*, 1992, **257**, 401.

55 N. Kim, K.M. Azadzoi, I. Goldstein and I.S. de Tejada, *J. Clin. Invest.*, 1991, **88**, 112.

56 F. Holmquist, H. Hedlund and K.-E. Andersson, *Acta Physiol. Scand.*, 1991, **141**, 441.

57 A.M. Franchi, M. Chaud, V. Rettori, A. Suburo, S.M. McCann and M. Gimeno, *Proc. Natl. Acad. Sci. USA*, 1994, **91**, 539.

58 I.S. de Tejada, I. Goldstein, K. Azadzoi, R.J. Krane and R.A. Cohen, *N. Eng. J. Med.*, 1989, **320**, 1025.
59 J. Garthwaite, S.L. Charles and R. Chess-Williams, *Nature*, 1988, **336**, 385.
60 K.F. Kitto, J.E. Haley and G.L. Wilcox, *Neurosci. Lett.*, 1992, **148**, 1.
61 S.T. Meller, P.S. Pechman, G.F. Gebhart and T.J. Maves, *Neuroscience*, 1992, **50**, 7.
62 Y.A. Kolensnikov, C.G. Pick and G.W. Pasternak, *Eur. J. Pharmacol.*, 1992, **221**, 399.
63 L.N. Shapoval, V.F. Sagach and L.S. Pobegailo, *Neurosci. Lett.*, 1991, **132**, 47.
64 H. Togashi, I. Sakuma, M. Yoshioka, T. Kobayashi, H. Ysuda, A. Kitabatake, H. Saito, S.S. Gross and R. Levi, *J. Pharmacol. Exptl. Ther.*, 1992, **262**, 343.
65 A.G.B. Kovách, C. Szabó, Z. Benyó, C. Csáki, J.H. Greenberg and M. Reivich, *J. Physiol.*, 1992, **449**, 183.
66 C. Iadecola, *Proc. Natl, Acad. Sci. USA*, 1992, **89**, 3913.
67 K.M. Boje and P.K. Arora, *Brain Res.*, 1992, **58**, 250.
68 J.P. Nowicki, D. Duval, H. Poignet and B. Scatton, *Eur. J. Pharmacol.*, 1991, **204**, 339.
69 S. Yamamoto, E.V. Golanov, S.B. Berger and D.J. Reis, *J. Cereb. Blood Flow Metab.*, 1992, **12**, 717.
70 S.A. Lipton, Y.-B. Choi, Z.-H. Pan, S.Z. Lei, H.-S.V. Chen, N.J. Sucher, J. Loscalzo, D.J. Singel and J.S. Stamler, *Nature*, 1993, **364**, 626.
71 B. Halliwell, *Lancet*, 1994, **344**, 721.
72 L.C. Green, D.A. Wagner, J. Glogowski, P.L. Skipper, J.S. Wishnok and S.R. Tannenbaum, *Anal Biochem.*, 1982, **126**, 131.
73 R. Iyengar, D.J. Stuehr and M.A. Marletta, *Proc. Natl. Acad. Sci. USA*, 1987, **84**, 6369.
74 J.B. Hibbs, Z. Vavrin and R.R. Taintor, *J. Immunol.*, 1987, **138**, 550.
75 X. Cui, C.L. Joannou, M.N. Hughes and R. Cammack, *FEMS Microbiol. Lett.*, 1992, **98**, 67.
76 N.V. Blough and O.C. Zafiriou, *Inorg. Chem.*, 1985, **24**, 3502.
77 R. Radi, J.S. Beckman, K.M. Bush and B.A. Freeman, *J. Biol. Chem.*, 1991, **266**, 4244.
78 J.S. Beckman, T.W. Beckman, J. Chen, P.A. Marshall and B.A. Freeman, *Proc. Natl. Acad. Sci. USA*, 1990, **87**, 1620.
79 J. Assruey, F.Q. Cunha, M. Epperlein, A. Noronha-Dutra, C.A. O'Donnell, F.Y. Liew and S. Moncada, *Eur. J. Immunol.*, 1994, **24**, 672.
80 W.H. Koppenol, J.J. Moreno, W.A. Pryor, H. Ischiropoulos and J.S. Beckman, *Chem. Res. Toxicol.*, 1992, **5**, 834.
81 M.A. Moro, V.M. Darley-Usmar, D.A. Goodwin, N.G. Read, R. Zamora-Pino, M. Feelisch, M.W. Radomski and S. Moncada, *Proc. Natl. Acad. Sci. USA*, 1994, **91**, 6702.
82 A.R. Butler and L.M. Renton, unpublished observations.
83 J.R. Verner, *Postgrad. Med. J.*, 1974, **50**, 576.
84 F.W. Flitney, I.L. Megson, D.E. Flitney and A.R. Butler, *Brit. J. Pharmacol.*, 1992, **107**, 842.
85 A.R. Butler and C. Glidewell, *Chem. Soc. Rev.*, 1987, **16**, 361.
86 E.A. Kowaluk, P. Seth and H.-L. Fung, *J. Pharmacol. Exptl. Ther.*, 1992, **262**, 916.

CHAPTER 5

Therapeutic Aspects of Manganese(II)-based Superoxide Dismutase Mimics

RANDY H. WEISS AND DENNIS P. RILEY

Monsanto Company, 800 N. Lindbergh Blvd., St. Louis, MO 63167, USA

1 Introduction

Since the discovery of the functional activity of the enzyme superoxide dismutase (SOD) by McCord and Fridovich,[1] intensive efforts have been made to develop the enzyme as a therapeutic agent for the treatment of a wide range of diseases and disorders, including reperfusion injury to the ischemic myocardium, rheumatoid arthritis, osteoarthritis, bronchopulmonary dysplasia, inflammatory bowel disease, post-traumatic and post-ischemic neuropathies, organ transplantation, and radiation-induced injury.[2,3] Although almost 30 years have passed since its discovery, the only clinically approved use for SOD has been in Europe for the treatment of osteoarthritis. Recently, encouraging results in clinical trials have been obtained with polyethylene glycol (PEG)-conjugated Cu,Zn SOD for the treatment of severe head injuries[4] and with recombinant human SOD for the prevention of multiple organ failure after multiple trauma.[5] However, the results of clinical trials to date with the SOD enzyme for the treatment of myocardial ischemia-reperfusion injury have been disappointing.[6–8] Another surge of interest in SOD developed with the discovery that there is a defective gene that codes for SOD in one inherited form of Lou Gehrig's disease (ALS, amyotrophic lateral sclerosis).[9,10]

The enzyme superoxide dismutase is an oxidoreductase that catalyses the dismutation of superoxide to hydrogen peroxide and molecular oxygen (equations 1 and 2).[11] In the absence of the enzyme, superoxide undergoes a slow self-dismutation process (equation 3). The role of the enzyme is to regulate the lifetime of superoxide and protect against superoxide-mediated cytotoxicity.

$$Mn(\text{II}) + HO_2^{\cdot} \rightarrow Mn(\text{III}) + H_2O_2 \qquad (1)$$

$$O_2^{-}\cdot + Mn(III) \rightarrow Mn(II) + O_2 \tag{2}$$

$$O_2^{-}\cdot + HO_2\cdot \xrightarrow{H^+} H_2O_2 + O_2 \tag{3}$$

The lack of success in the development of the SOD enzyme as a therapeutic agent can be attributed to several factors: (1) a narrow efficacious dose characterized by a bell-shaped dose–response curve; (2) lack of an accurate and routine method to monitor and quantitate SOD activity; (3) inability of the enzyme to gain access to the intracellular space; (4) instability of the enzyme; (5) lack of oral bioavailability; (6) immunogenicity; and (7) cost. We felt that many of the limiting factors could be overcome by the development of synthetic, low-molecular-weight mimetics of the SOD enzyme and by use of a direct method to measure the SOD activity of these complexes.

Our goal has been to develop the Mn(II)-based SOD mimics shown in Figure 1 as therapeutic agents for diseases mediated by superoxide, particularly for the treatment of myocardial ischemia-reperfusion injury. We will discuss how the Mn(II)-based SOD mimics were designed and illustrate the advantages of these SOD mimics over the SOD enzymes with regard to their SOD activity, stability, pharmacology and cost.

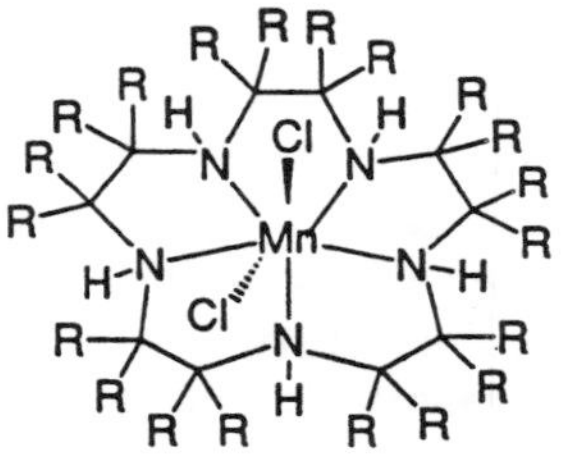

Figure 1 *Structural class of SOD mimics based upon the Mn(II) dichloro complex of 1,4,7,10,13-pentaazacyclopentadecane*

2   Results and Discussion

Therapeutic Strategy of Superoxide Dismutase Mimics

The basis of the therapeutic strategy utilizing SOD mimics (see Figure 2) is to more efficiently catalyze the dismutation of superoxide to generate the less innocuous non-radical compounds, hydrogen peroxide and molecular oxygen, before cytotoxic superoxide-derived oxidants, such as peroxynitrite and the perhydroxyl radical can be produced. Superoxide reacts in a near diffusion-controlled manner with nitric oxide to produce the potent oxidant peroxynitrite.[12] Under ischemic conditions where the pH of tissue can decrease, the formation of the perhydroxyl radical $(HO_2\cdot)$, which can abstract hydrogen atoms from unsaturated fatty acids to initiate lipid peroxidation,[13] will occur to a greater extent. By *catalyzing* the dismutation of superoxide, the formation of these deleterious oxidants can be inhibited without consumption of the SOD mimic. This mechanism is in contrast to the mechanism of action of a general

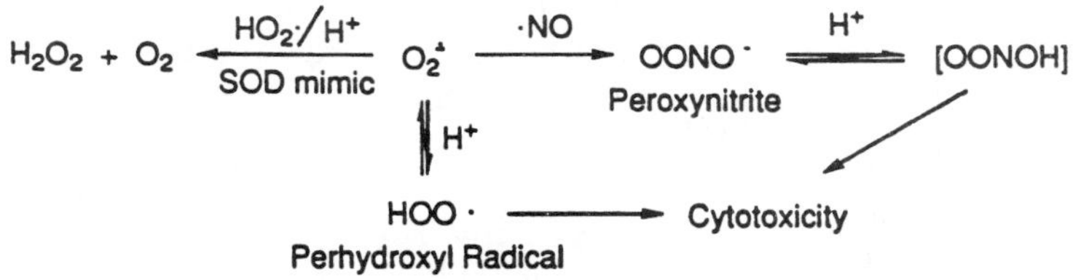

Figure 2 *Therapeutic strategy for the SOD mimics. The SOD mimic catalyzes the dismutation of superoxide, thereby preventing the formation of the cytotoxic oxidants, peroxynitrite and the perhydroxyl radical*

antioxidant, which would be consumed by reacting *stoichiometrically* with the oxidant or with a radical derived from the oxidant.

Stopped Flow Kinetic Analysis: A Direct Assay for Superoxide Dismutase Activity

Numerous indirect assays, such as the cytochrome c assay, have been used in attempts to measure the SOD activity of putative SOD mimics.[14-16] However, these assays, which typically rely on a spectrophotometric change of a redox indicator to measure superoxide levels, cannot kinetically distinguish between a catalytic dismutation of superoxide and a stoichiometric interaction of superoxide with the putative SOD mimic. Moreover, the indirect assays are prone to false positives or false negatives respectively, when the putative SOD mimic oxidizes or reduces the spectrophotometric indicator.

We recognized the need for methodology to measure SOD activity directly that would be more accessible to the bench-top scientist than is the method of pulse radiolysis, another direct measure. Consequently, we developed methodology to measure the catalytic dismutation of superoxide by stopped-flow kinetic analysis.[17,18] By this technique, we directly monitor the decay of superoxide spectrophotometrically in the presence or absence of a putative SOD mimic at a given pH. Kinetic analysis of this decay can determine whether the complex is a SOD mimic (decay of superoxide becomes first-order in superoxide and first-order in complex; see equations 1 and 2), or is inactive (decay of superoxide remains second-order for its self-dismutation; see equation 3). At least a tenfold excess of superoxide over the putative SOD mimic is used in the stopped-flow assay, to eliminate contributions due to a stoichiometric reaction of the complex with superoxide. A catalytic rate constant (k_{cat}) for the dismutation of superoxide by the complex can be determined from the observed rate constants of superoxide decay as a function of catalyst concentration.[17,18]

Design of Manganese-based Superoxide Dismutase Mimics

Manganese-based complexes are particularly attractive as potential SOD mimics because they have a reduced capacity to react with the dismutation product hydrogen peroxide to generate cytotoxic hydroxyl radicals (Fenton chemistry), whereas copper- and iron-based complexes and their aquated ions are known to carry out such undesirable chemistry.[19–23] Early work by many researchers identified a number of manganese-based complexes that had apparent SOD activity by indirect assays.[24–27] However, when we assessed the SOD activity of these complexes by the direct method of stopped-flow kinetic analysis, we observed no SOD activity with aquo Mn(II) and manganese complexes of desferal,[18] cyclam,[28] 8-quinolinol,[29] or salen.[29] The apparent 'SOD activity' of the above manganese complexes is due to a stoichiometric, not catalytic, reaction of the complex with superoxide.[18] It is absolutely critical that the assay used for measuring SOD activity of putative SOD mimics be able to distinguish between a *catalytic* dismutation of superoxide and a *stoichiometric* reaction with superoxide. Additionally, it is apparent that such an assay must also be capable of rapidly and reliably providing quantitative rate data so that useful structure–activity relationships can be established for the optimization/design of such synthetic catalysts.

We proposed that a Mn(II) complex of a macrocyclic polyamine could have SOD activity, and also have the desirable characteristic of high stability due to the enhanced kinetic stabilizing effect of the macrocycle.[30] To this end, we synthesized over twenty different Mn(II)-based macrocyclic amines with macrocycles differing in size (9–21 membered rings), number of donor nitrogen atoms (3–7 nitrogens), type of heteroatoms (nitrogen replaced by oxygen), and degree of unsaturation within the ring. The SOD activity of each complex was assessed by stopped-flow kinetic analysis. Of these complexes, only the Mn(II) complex of 1,4,7,10,13-pentaazacyclopentadecane (Figure 1; R = H) had significant SOD activity, with a catalytic rate constant (k_{cat}) of $4.13 \times 10^7 \, M^{-1} s^{-1}$ at physiological pH.[28] The crystal structure of this complex (Figure 3) indicates that the five nitrogen atoms are coordinated to the Mn(II) center, and largely in the plane of the Mn with the chloride ions coordinated in the *trans*-axial positions.[28]

The Mn(II)-based SOD mimics are high-spin, d^5 complexes, and in aqueous solution are univalent cations, as shown by conductivity measurements.[28] The Mn(II) complexes have high oxidation potentials (> 0.7 V *vs.* NHE). The high oxidative stability of these complexes is significant in that there are likely to be very few biological species with sufficient oxidizing power to convert the Mn(II) complex into the Mn(III) complex; although oxidants such as the perhydroxyl radical ($HO_2\cdot$) and the hydroxyl radical ($HO\cdot$) are indeed competent for this oxidation. The Mn(II) complexes of this pentaaza crown family are white crystalline solids synthesized by the addition of the macrocyclic ligand to a methanolic solution of anhydrous $MnCl_2$.[28] We have developed a number of versatile, asymmetric syntheses of highly functionalized polyazamacrocycles *via* the reduction of cyclic peptide precursors and via a bis(chloroacetamide) approach.[31,32] As will be discussed later, Mn(II)-based complexes with stereochemically-defined substituents can be prepared so that such substituents enhance both the SOD activity and stability of these complexes.

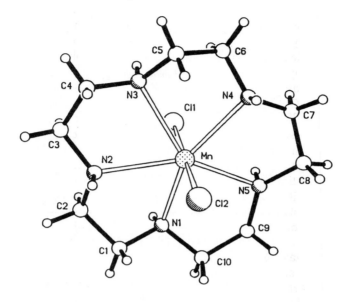

Selected Bond Lengths		Selected Bond Angles(°)	
Bond	Length (Å)	Type	Angle(°)
Mn-Cl$_1$	2.635(4)	ClMnCl	178.9(2)
Mn-Cl$_2$	2.571(4)	N$_1$MnN$_5$	79.1(6)
Mn-N$_1$	2.26(1)	N$_1$MnN$_2$	71.6(6)
Mn-N$_2$	2.41(1)	N$_2$MnN$_3$	67.6(6)
Mn-N$_3$	2.45(1)	N$_3$MnN$_4$	71.9(6)
Mn-N$_4$	2.40(1)	N$_4$MnN$_5$	74.2(6)
Mn-N$_5$	2.27(2)	Cl$_1$MnN$_1$	84.2(4)
		Cl$_1$MnN$_2$	97.2(4)
		Cl$_1$MnN$_3$	78.0(4)
		Cl$_1$MnN$_4$	92.7(4)
		Cl$_1$MnN$_5$	86.5(5)

Figure 3 *Crystal structure of the SOD mimic SC-52608*

Advantages of Superoxide Dismutase Mimics over Superoxide Dismutase Enzymes

The Mn(II)-based SOD mimics have numerous potential advantages over the SOD enzymes as potential therapeutic agents, including membrane permeability, selective reactivity for superoxide, immunogenicity effects, stability, and cost. Table 1 compares characteristics of the Mn(II)-based SOD mimics with R = H *versus* those of bovine erythrocyte Cu,Zn SOD.

Molecular Weight. The low molecular weight of the SOD mimics, lower than SOD enzyme by a factor of almost 100, permits the complexes to gain access to

Table 1 *Advantages of Mn(II)-based SOD mimics over SOD enzymes*

Characteristic		Bovine erythrocyte Cu, Zn SOD enzyme	Advantage of SOD mimic
Molecular weight	341	32 500	Intracellular
Reactivity	$O_2^-\cdot$ only	O_2^-, $OONO^-$, H_2O_2	Low toxicity
Structure	Non-peptidic	Peptide	Oral bioavailability; non-immunogenicity
Kinetic stability with EDTA	Stable	Unstable	Stability
Source	Synthesis	Isolation	Cost

the intracellular space of certain cell types,[33] whereas the cells are generally impermeable to the enzyme.

Reactivity. The SOD mimics catalyze the decomposition of superoxide only and do not react with hydrogen peroxide or with peroxynitrite. The Cu,Zn SOD enzyme reacts with both hydrogen peroxide[34] and with peroxynitrite,[35] inactivating the enzyme. The selective reactivity of the SOD mimics for superoxide makes these complexes useful probes to establish the role of superoxide in biological mechanisms.

Structure. The SOD enzyme structure is peptide-based and is thus readily degraded by proteases; therefore, it is not orally bioavailable. Immunogenicity problems are also a concern with the non-human-derived SOD enzymes, whereas the SOD mimics, which are not peptide-based, do not have that problem. It should also be possible to develop an orally effective Mn(II)-based SOD mimic.

Kinetic Stability. Due to the stabilizing effect of the macrocycle, the SOD mimics have high kinetic stability (discussed below) and, for example, are stable in the presence of the transition metal ion chelators, such as EDTA. However, the apoenzyme form of the Cu,Zn SOD enzyme can readily be prepared by dialyzing the enzyme with EDTA.[36,37]

Source. The Mn(II)-based SOD mimics are prepared by total synthesis, whereas the enzymes are isolated from natural sources or prepared by recombinant DNA techniques. Consequently, the cost of manufacturing the SOD mimics would be potentially much less than that for manufacturing the enzymes.

Conformational Control of Superoxide Dismutase Activity and Stability

From extensive studies of the structure-activity relationships, we have observed that the conformation of substituents on the macrocyclic backbone of the

Mn(II)-based SOD mimics is critical in determining the complex's SOD activity and stability. We have identified the *trans*-2,3-cyclohexano group as a substituent that increases both the SOD activity[37–39] and kinetic stability of the Mn complex. Figure 4 shows the SOD activities and kinetic stabilities (defined by the second-order rate constant k_{diss} (first-order in [H^+] and first-order in [Mn(L)] complex) of the unsubstituted SOD mimic SC-52608, the 2,3-(*R,R*)-*trans*-cyclohexano substituted SC-54417,[37] and the 2,3-(*R,R*)-8,9-(*R,R*)-bis(*trans*-cyclohexano) substituted SC-55858.[39] The k_{cat} values for the dismutation of superoxide by these complexes increase as additional *trans*-cyclohexano groups are added, with SC-55858 ($k_{cat} = 1.20 \times 10^8 \, M^{-1} s^{-1}$) having three times more SOD activity that SC-52608 ($k_{cat} = 4.13 \times 10^7 \, M^{-1} s^{-1}$) at pH 7.4.[39]

By analysis of the kinetics of superoxide decay, the mechanism by which these Mn(II) complexes catalyze the dismutation of superoxide has been studied.[28,37,39] The kinetics of superoxide decay for all three complexes, SC-52608, SC-54417, and SC-55858, shows in each case that the superoxide decay rates are first-order in both [Mn(L)] and superoxide concentrations. Results of mechanistic studies and kinetic studies indicate that the rate-determining step in the catalytic cycle involves oxidation of the Mn(II) to Mn(III). Further, the rate of superoxide decay was studied at different pH values, and we observed that the k_{cat} value increases in a linear fashion as the proton concentration increases, although the kinetics are less than first-order in proton concentration. Consequently, two rate-determining steps are implied: a step that is dependent on the concentration of protons, and one that is independent of pH. The magnitude of the relative contribution of each pathway is different with each Mn(II) complex. The proton-dependent pathway is consistent with the oxidation of the Mn(II)-based complex by the perhydroxyl radical *via* an outer-sphere mechanism. The proton-independent pathway is consistent with the oxidation of the Mn(II) complex by an inner-sphere pathway involving binding of the superoxide anion to a vacant coordination site on Mn(II). This mechanistic interpretation is also consistent with deuteration results in which an isotope effect of nearly 6 is observed for the proton-dependent step. Additionally, the rate of the pH-independent path in D_2O is affected indicating that water exchange on the Mn(II) center is involved.

The magnitude of the rate constants measured for these two pathways, nearly diffusion controlled for the pH-dependent path and near $1 \times 10^{+7} s^{-1}$ for the pH-independent path, are significant. The exchange rate of water on Mn(II) generating a vacant coordination site is on the order of $10^{+7} s^{-1}$, similar to the rate for the inner-sphere process measured for these complexes. The near diffusion controlled rate for subsequent electron transfer by this path and the proton coupled electron transfer of the outer-sphere path dictate that the structure of the oxidized and reduced forms of the active catalytic species be similar in structure. We have proposed that this can be accomplished by folding of the ligand to generate a pseudo-octahedral six-coordinate geometry about the high-spin d^5 Mn(II) ion. Thus, if the ligand has the proper substitution pattern, *i.e.*, substituents arranged in a favorable stereochemistry so that the ligand is forced by steric repulsive forces to fold about the spherically symmetrical Mn(II) ion, then the ligand will force a geometry about the Mn(II) complex which can have high catalytic SOD activity. Thus, the substituents on the ligand governs the electron transfer process by affecting the degree of folding of the macrocyclic ligand on Mn(II). Consistent with this is that complexes derived from ligands in this series which modeling shows are rigidly planar do not, in fact, possess catalytic activity.[39]

Substituents on the macrocyclic ligand exert major effects on the stability as well as on the catalytic activity. The thermodynamic stabilities ($\log K$) of the Mn(II) complexes were assessed by potentiometric titration. As can be seen from Figure 4, the Mn(II) complexes are very thermodynamically stable, with $\log K$ values greater than 10.8. The thermodynamic stability of the complexes can be substantially increased by the addition of the *trans*-cyclohexano substituent, which is believed to increase the preorganizational rigidity of the Mn(II)-based complexes similarly to that observed with *gem*-dimethyl substituents on macro- cyclic tetrathioether complexes of Ni(II).[40] For example, the mono-substituted SC-54417 is more thermodynamically stable than the unsubstituted SC-52608, by a factor of almost 10. For SC-55858, with two *trans*-cyclohexano substituents, the corresponding Mn(II) complex is over 200 times more thermodynamically stable than the unsubstituted SC-52608.

The mechanism of ligand loss for these Mn(II) complexes has been studied[37] and is a second-order process, first-order in [H^+] and first-order in [Mn(L)]

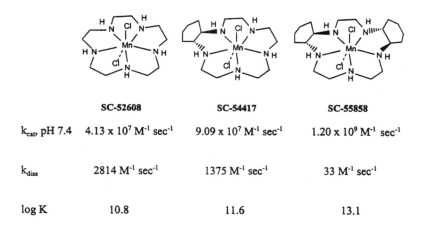

	SC-52608	SC-54417	SC-55858
k_{cat}, pH 7.4	4.13×10^7 M^{-1} sec^{-1}	9.09×10^7 M^{-1} sec^{-1}	1.20×10^9 M^{-1} sec^{-1}
k_{diss}	2814 M^{-1} sec^{-1}	1375 M^{-1} sec^{-1}	33 M^{-1} sec^{-1}
$\log K$	10.8	11.6	13.1

Increasing SOD Activity and Stability

⟹

Figure 4 *Catalytic rate constants (k_{cat}) for the dismutation of superoxide and the kinetic stabilities (k_{diss}, the dissociative rate constant) of the Mn(II)-based SOD mimics SC-52608, SC-54417, and SC-55858. The k_{cat} values are measured by stopped-flow kinetic analysis, and the k_{diss} values are measured by spectrophotometrically monitoring the exchange of Cu(II) for the Mn(II) in the complex*

complex, but zero-order in added metal ions or added chelating ligands, such as EDTA. This permits a facile spectrophotometric kinetic assay in which the kinetic stability of the complexes is measured by the exchange rate of Cu(II) for Mn(II) in the complex (equation 4, where L = ligand).[37,41] The second-order rate constant, designated as k_{diss}, also decreases with increasing numbers of *trans*-cyclohexano groups reflecting the increased kinetic stability. The rate of exchange is pH-dependent and independent of the Cu(II) concentration.

$$Mn(L)^{2+} + Cu^{2+} \rightarrow Mn^{2+} + Cu(L)^{2+} \tag{4}$$

The kinetic stability of the mono-*trans*-cyclohexano substituent SC-54417 is twice that of the unsubstituted SC-52608, and two *trans*-cyclohexano groups increase the kinetic stability of the Mn complex SC-55858 to over 85 times that of the unsubstituted complex SC-52608. The increased kinetic stability of the Mn complexes by the addition of *trans*-cyclohexano groups is believed to be due to the greater preorganizational rigidity of the complexes. In general we observe that increasing the number of substituents increases the kinetic and thermodynamic stability of the resultant Mn(II) complexes.

Inhibition of Neutrophil-Mediated Human Aortic Endothelial Cell Injury by Superoxide Dismutase Mimics

We have reported the observation that the Mn(II)-based SOD mimics inhibit superoxide-mediated injury to human aortic endothelial cells.[38] The amount of injury to human aortic endothelial cells was assessed by pre-labeling the cells with [^{51}Cr]chromate, which is released into the medium when the cells are injured. Superoxide was produced either by activated human neutrophils or by xanthine/xanthine oxidase. The amount of injury to cells was directly correlated to the amount of superoxide the cells were exposed to.

Bovine erythrocyte Cu,Zn SOD protected against the superoxide-mediated injury to the endothelial cells, but the results were quite variable, and a bell-shaped dose–response curve was observed. However, the SOD mimics SC-52608 and SC-54417 showed reproducible dose-dependent protection against the injury, with nearly complete inhibition being obtained with $150\,\mu M$ complex. SC-54385, a Mn(II) complex that has no detectable SOD activity, did not protect the cells against the superoxide-mediated injury, even at concentrations as high as $300\,\mu M$. These results are consistent with human neutrophil-mediated injury to aortic endothelial cells being mediated by superoxide.

Lipophilicity

Clearly, a potential advantage of the Mn(II)-based SOD mimics over the SOD enzymes is the ability of the SOD mimics to permeate the cell membrane.[33] The lipophilicity of the complexes can be controlled by placing the appropriate substituents onto the carbon backbone of the macrocycle. As shown in Table 2, the lipophilicities of mono-substituted Mn(II)-based complexes can be varied

Table 2 *Lipophilicities of Mn(II)-based SOD mimics*

R	log P
1-Aminobutyl	− 4.1
Hydroxymethyl	− 3.4
H	− 2.9
Methyl	− 2.6
2-Propynyl	− 2.5
Isobutyl	− 1.5
Benzyl	− 1.3
Alkenyl	− 1.3
Phenyl	− 1.2
Cyclohexyl	− 0.87
Cyclohexylmethyl	− 0.30
1-Naphthyl	+ 0.02
n-Octadecyl	+ 0.18

P is the ratio of the partitioning of the complex between *n*-octanol and aqueous Hepes buffer, pH 7.4.

over a factor of 1000, as assessed by log *P* (*P* = the ratio of the partitioning of the complex between *n*-octanol and aqueous Hepes buffer, respectively). The unsubstituted complex (R = H in Table 2) has a log *P of* − 2.9, *indicating that the complex is about* 1000 *times more soluble in aqueous buffer than in the n-octanol phase*. Depending on the substituent, a complex can be prepared that is either more or less lipophilic than the unsubstituted complex. For example, the 4-aminobutyl derivative is less lipophilic (more negative log *P*) than is the unsubstituted complex. The *n*-octadecyl derivative [R = CH$_2$(CH$_2$)$_{16}$CH$_3$ in Table 2] has a positive log *P* of 0.18, indicating that the complex is slightly more soluble in the *n*-octanol phase than in the aqueous phase. Of particular interest is the ability to increase the lipophilicity of the complexes with the *trans*-cyclohexano substituent, which also increases the SOD activity and the kinetic and thermodynamic stability of the complexes. SC-55858, which has two *trans*-cyclohexano substituents, has a log *P* of − 0.63, whereas SC-54417, with one cyclohexano group, has a log *P* of − 1.9.

In the neutrophil-mediated, endothelial cell injury assay, the SOD mimics protected against the injury to a greater extent than the SOD enzyme did under conditions where the same amounts of SOD activity of the SOD mimics and of the enzyme were used.[38] These results are consistent with the SOD mimics gaining access to the intracellular space of the cells, thereby catalyzing the dismutation of intracellular superoxide, which the SOD enzyme cannot gain access to. Therefore, the lipophilicities of the SOD mimics are expected to be critical in obtaining optimal protection of the intracellular space against superoxide-mediated cytotoxicity.

Inhibition of Myocardial Ischemia-Reperfusion Injury in the Isolated Heart

The protective effects of the Mn(II)-based SOD mimics against myocardial ischemia/reperfusion injury in the isolated rabbit heart and monkey heart have been investigated. Kilgore *et al.*[42] subjected Langendorff perfused, isolated rabbit hearts to a 30 minute period of global ischemia followed by a 45 minute period of reperfusion. Upon reperfusion, an increase in left ventricular end-diastolic pressure occurred, which was attenuated when the hearts were perfused with 20 μM SC-52608. Perfusion of SC-52608 also inhibited the release of creatine kinase and intracellular potassium, and reduced the extent of antibody binding to the intracellular protein myosin which occurs upon reperfusion of the ischemic heart. These studies indicated that SC-52608 is cardioprotective to the reperfused, ischemic isolated rabbit heart.

Friedrichs *et al.*[43] have also demonstrated a protective effect of SC-52608 from ischemia-reperfusion injury in the isolated primate heart. The isolated hearts were subjected to 35 minutes of global ischemia followed by 45 minutes of reperfusion. Perfusion of 48 μM SC-52608 attenuated the increase in end-diastolic pressure and the increase in creatine kinase release following reperfusion. The results indicate that the SOD mimics provide significant protection against myocardial ischemia/reperfusion injury.

Protection against Myocardial Ischemia-Reperfusion Injury *In Vivo*

The SOD enzyme has been reported to protect against myocardial reperfusion to ischemic tissue;[44] however, there are also many reports in the literature that indicate the enzyme does not offer protection.[45] In those cases where protection has been observed and dose responses have been examined, the SOD enzyme shows a bell-shaped dose–response curve. The discrepancy among reports of the effectiveness of the SOD enzyme may be due to a narrow bell-shaped dose–response curve, different doses of the enzyme, variable periods of ischemia and reperfusion and short half-lives of the SOD enzymes.[44,45]

Black *et al.*[46] have examined the protective effect of the SOD mimic SC-52608 in an *in vivo* dog model of myocardial ischemia-reperfusion injury. The left circumflex coronary artery of the anesthetized dog was occluded for 90 minutes, and an 18 hour period of reperfusion followed. Intravenous administration of the SOD mimic SC-52608 at a total dose of 16 mg kg^{-1} significantly reduced the myocardial infarct size observed in untreated dogs. However, a Mn(II) complex with no SOD activity, SC-54385, at the same dose failed to protect against the myocardial injury. These results are consistent with superoxide mediating the myocardial ischemia-reperfusion injury; they illustrate the potential usefulness of the Mn(II)-based SOD mimics in protecting against such injuries *in vivo*.

We have investigated the protective effects of the Mn(II)-based SOD mimics in a feline model of myocardial ischemia-reperfusion injury.[47] In this *in vivo* model, the hearts of the animals were made ischemic for a period of 75 minutes, which was followed by a reperfusion period of 4.5 hours. At that time the animals were sacrificed and the degree of necrosis and the area at risk in the myocardium were determined. In the feline model, the bovine erythrocyte Cu,Zn SOD enzyme showed a bell-shaped dose–response curve with significant protection against necrosis only at an intravenous dose of $16 \, mg \, kg^{-1}$. However, the SOD mimic SC-54417 inhibited the necrosis in a dose-dependent manner, with an effective dose being observed at $5 \, \mu mol \, kg^{-1}$ intravenously. Even at a dose as high as $50 \, \mu mol \, kg^{-1}$, significant protection by SC-54417 was observed against the reperfusion injury, indicating that there is a difference between the protective effects of the SOD mimic and those of the enzyme, as the mimic does not exhibit a bell-shaped dose–response curve.

Further structure–activity relationship studies have been carried out *in vivo* to assess the role of superoxide in myocardial ischemia-reperfusion injury. The Mn(II)-based SOD mimic SC-55858 also protected against the reperfusion injury in the feline model at an intravenous dose of $5 \, \mu mol \, kg^{-1}$. A Mn(II)-based complex with no SOD activity did not protect at an intravenous dose of $5 \, \mu mol \, kg^{-1}$, suggesting that superoxide is a mediator of the reperfusion injury to the ischemic myocardial tissue.[47] Furthermore, direct administration of $MnCl_2$ did not inhibit the formation of the necrotic tissue in this model, which indicates that the protection observed with the Mn(II)-based SOD mimics is due to the intact complex.

Potentiation of Nitric Oxide

We evaluated whether the Mn(II)-based SOD mimics could potentiate the levels of nitric oxide, a potent vasorelaxant.[12] Superoxide reacts with nitric oxide in a diffusion-controlled manner to produce peroxynitrite. By catalyzing the dismutation of superoxide, SOD mimics would be expected to increase nitric oxide levels. The SOD mimic SC-52608 enhanced nitric oxide levels (as assessed by cyclic GMP activity) in rat lung fibroblasts in a dose-dependent manner.[48] SC-52608 induced the relaxation of preconstricted rat aortic rings.[48] The aortic ring relaxation was endothelium-dependent and inhibitable by a nitric oxide synthase inhibitor. Intravenous administration of SC-52608 into conscious rats resulted in a transient, dose-dependent decrease in blood pressure.[48] The results are consistent with the SOD mimic SC-52608 potentiating levels of nitric oxide, which causes the observed relaxation of the aortic rings and the decrease in blood pressure.

Anti-Inflammatory Activities of Manganese(II)-based Superoxide Mimics

Superoxide is a product of activated polymorphonuclear leukocytes, such as the neutrophils, and has been proposed to be a mediator of inflammation.[49,50] We have evaluated the role of superoxide in acetic acid-induced, neutrophil-dependent inflammation of the colon in mice.[51] The inflammation of the colonic tissue was evaluated biochemically for the neutrophil marker enzyme, myeloperoxidase, and also histologically. Colonic myeloperoxidase activity correlates with the degree of tissue inflammation as assessed by visual and histological analysis. The Mn(II)-based SOD mimic SC-52608, when administered intracolonically, inhibited the acetic acid-induced inflammation of the colon in a dose-dependent manner (ED_{50} for SC-52608 is $10\,mg\,kg^{-1}$). When administered intravenously at a dose of $10\,mg\,kg^{-1}$, SC-52608 inhibited the acetic acid-induced inflammation by 50%, as assessed by tissue myeloperoxidase activity.

Monosubstituted Mn(II)-based complexes, which catalyze the dismutation of superoxide as shown by stopped-flow kinetic analysis, were tested for anti-inflammatory activity in the mouse acetic acid-induced colitis model. As can be seen from Table 3, all of the SOD mimics are anti-inflammatory. Histological analysis of the colonic tissue confirmed these results. Of particular importance is the observation that Mn(II) complexes that have no SOD activity, specifically the Mn(II) dichloro complexes of 1,4,7,10,13-pentaazacyclohexadecane and 1,4,7,11,14-pentaazacycloheptadecane, do not protect against the colonic inflammation induced by acetic acid when the compounds are administered intracolonically at a dose of $30\,mg\,kg^{-1}$. These results are consistent with a role for superoxide as a mediator of neutrophil-dependent inflammation. Consistent with this hypothesis is the observation that SC-52608 inhibits neutrophil-dependent inflammation induced by the intradermal administration of leukotriene B_4, a neutrophil chemoattractant.[51]

As shown by histological analysis, SC-52608 and the other SOD mimics attenuate the acetic acid-induced colitis in mice by inhibiting the influx of neutrophils. A possible mechanism for inhibition of the neutrophil infiltration

Table 3 *Anti-inflammatory activity of Mn(II)-based SOD mimics*

R	X	% Inhibition MPO activity[a] ($n =$)
H	Cl	69
Methyl	Cl	47
Isobutyl	Cl	70
Phenylmethyl	Cl	48
Cyclohexylmethyl	Cl	77
H	OAc	32

[a]A solution of 3% acetic acid was instilled intracolonically into mice. After 24 hours, the mice were killed and colonic tissue samples were collected. Tissue samples were assessed biochemically for the neutrophil marker enzyme myeloperoxidase (MPO). Mn(II) complexes were given intra-colonically 30 minutes prior to acetic acid.

is that the SOD mimics effectively catalyze the dismutation of superoxide, thereby potentiating nitric oxide levels. Nitric oxide inhibits the adhesion of neutrophils to endothelial cells, which prevents the migration of the inflammatory cells.[52-54]

3   Summary

Mn(II)-based complexes of the macrocyclic pentaamine 1,4,7,10,13-pentaazacyclopentadecane catalyze the dismutation of superoxide and protect against superoxide-mediated tissue injury both *in vitro* and *in vivo*. The SOD mimics are currently undergoing pre-clinical studies as potential therapeutic agents for treating myocardial ischemia-reperfusion injury. The complexes are also anti-inflammatory and potentiate nitric oxide levels by inhibiting production of peroxynitrite. The Mn(II)-based SOD mimics are useful as probes to elucidate biological mechanisms that may be mediated by superoxide.

Acknowledgments

We wish to thank everyone who has contributed positively to this project, including our colleagues at Monsanto Corporate Research and G. D. Searle Co., our consultants, and our external collaborators. We want to acknowlegde with special thanks the expert molecular pharmacology leadership of Drs. Mark Currie and Daniela Salvemini of the G. D. Searle Co.

References

1 J.M. McCord and I. Fridovich, *J. Biol. Chem.*, 1969, **244**, 6049.
2 A. Petkau, *Cancer Treat. Rev.*, 1986, **13**, 17.
3 A.M. Michelson, *Life Chem. Rep.*, 1987, **6**, 1.
4 J.P. Muizelaar, A. Marmarou, H.F. Young, S.C. Choi, A. Wolf, R.L. Schneider and H.A. Kontos, *J. Neurosurg.*, 1993, **78**, 375.
5 I. Marzi, V. Bühren, A. Schüttler and O. Trentz, *J. Trauma*, 1993, **35**, 110.
6 J.T. Flaherty, P. Bertram, J.W. Gruber, R.R. Heuser, D.A. Rothbaum, L.R. Burwell, B.S. George, D.J. Kereiakes, D. Deitchman, N. Gustafson, J.A. Brinker, L.C. Becker, J. Mancini, E. Topol and S.W. Werns, *Circulation*, 1994, **89**, 1982.
7 T. Shibata, F. Yamamoto, Y. Kosakai, K. Kawazoe, H. Komai, H. Ichikawa, T. Ohashi, Y. Shimada, N. Nakajima and Y. Kawashima, *Nippon Kyobu Geka Gakkai Zasshi.*, 1993, **41**, 427.
8 Y. Murohara, Y. Yui, R. Hattori and C. Kawai, *Am. J. Cardiol.*, 1991, **67**, 765.
9 H.-X. Deng, A. Hentati, J.A. Tainer, Z. Iqbal, A. Cayabyab, W.-Y. Hung, E.D. Getzoff, P. Hu, B. Herzfeldt, R.P. Roos, C. Warner, G. Deng, E. Soriano, C. Smyth, H.E. Parge, A. Ahmed, A.D. Roses, R.A. Hallewell, M.A. Pericak-Vance and T. Siddique, *Science*, 1993, **261**, 1047.
10 D.R. Rosen, T. Siddique, D. Patterson, D.A. Figlewicz, P. Sapp, A. Hentati, D. Donaldson, J. Goto, J.P. O'Regan, H.-X. Deng, Z. Rahman, A. Krizus, D. McKenna-Yasek, A. Cayabyab, S.M. Gaston, R. Berger, R.E. Tanzi, J.J. Halperin, B. Herzfeldt, R. Van den Bergh, W.-Y. Hung, T. Bird, G. Deng, D.W. Mulder, C. Smyth, N.G.

Laing, E. Soriano, M.A. Pericak-Vance, J. Haines, G.A. Rouleau, J.S. Gusella, H.R. Horvitz and R.H. Brown Jr., *Science*, 1993, **362**, 59.

11 I. Fridovich, *J. Biol. Chem.*, 1989, **264**, 7761.

12 J.S. Beckman and J.P. Crow, *Biochem. Soc. Trans.*, 1993, **21**, 330.

13 J. Aiken and T.A. Dix, *J. Biol. Chem.*, 1991, **266**, 15 091.

14 S. Goldstein and G. Czapski, *Free Radicals Res. Commun.*, 1991, **12–13**, 5.

15 S. Goldstein, C. Michel, W. Bors, M. Saran and G. Czapski, *Free Radicals Biol. Med.*, 1989, **4**, 295.

16 L. Flohe and F. Otting, *Methods Enzymol.*, 1984, **105**, 93.

17 D.P. Riley, W.J. Rivers and R.H. Weiss, *Anal. Biochem.*, 1991, **196**, 344.

18 R.H. Weiss, A.G. Flickinger, W.J. Rivers, M.M. Hardy, K.W. Aston, U.S. Ryan and D.P. Riley, *J. Biol. Chem.*, 1993, **268**, 23 049.

19 C. Walling and A. Goosen, *J. Am. Chem. Soc.*, 1973, **95**, 2987.

20 Y. Luo, K. Kustin and I.R. Epstein, *Inorg. Chem.*, 1988, **27**, 2489.

21 D.S. Sigman, D.R. Graham, V. D'Aurora and A.M. Stern, *J. Biol. Chem.*, 1979, **254**, 2269.

22 C. Walling, *Acc. Chem. Res.*, 1975, **8**, 125.

23 M. Masarwa, H. Cohen, D. Meyerstein, D.L. Hickman, A. Bakac and J.H. Espenson, *J. Am. Chem. Soc.*, 1988, **110**, 4293.

24 W.F. Beyer Jr. and I. Fridovich, *Arch. Biochem. Biophys.*, 1989, **271**, 149.

25 J.D. Rush, Z. Maskos and W.H. Koppenol, *Arch. Biochem. Biophys.*, 1991, **289**, 97.

26 J.K. Howie and D.T. Sawyer, *J. Am. Chem. Soc.*, 1976, **98**, 6698.

27 M. Baudry, S. Etienne, A. Bruce, M. Palucki, E. Jacobsen and B. Malfroy, *Biochem. Biophys. Res. Commun.*, 1993, **192**, 964.

28 D.P. Riley and R.H. Weiss, *J. Am. Chem. Soc.*, 1994, **116**, 387.

29 R.H. Weiss, A. Flickinger, W.J. Rivers, M.D. Hardy, K.W. Aston, U.S. Ryan, and D.P. Riley, *J. Biol. Chem.*, 1993, **268**, 23 049.

30 D.H. Busch, *Chem. Rev.*, 1993, **93**, 847.

31 K.W. Aston, S.L. Henke, A.S. Modak, D.P. Riley, K.R. Sample, R.H. Weiss and W.L. Neumann, *Tetrahedron Lett.*, 1994, **35**, 3687.

32 P.J. Lennon, H. Rahman, K.W. Aston, S.L. Henke and D.P. Riley, *Tetrahedron Lett.*, 1994, **35**, 853.

33 M.M. Hardy, M.E. Smith and C. Schasteen, unpublished results.

34 S.L. Jewett, S. Cushing, F. Gillespie, D. Smith and S. Sparks, *Eur. J. Biochem.*, 1989, **180**, 569.

35 H. Ischiropoulos, L. Zhu, J. Chen, M. Tsai, J.C. Martin, C.D. Smith and J.S. Beckman, *Arch. Biochem. Biophys.*, 1992, **298**, 431.

36 M.W. Pantoliano, J.S. Valentine, R.J. Mammone and D.M. Scholler, *J. Am. Chem. Soc.*, 1982, **104**, 1717.

37 D.P. Riley, S.L. Henke, P.J. Lennon, R.H. Weiss, W.L. Neumann, W. J. Rivers, K.W. Aston, K.R. Sample, H. Rahman, C.-S. Ling, J.-J. Shieh, D. Busch and W. Szulbinski, *Inorg. Chem.*, 1996, **35**, 5213.

38 M.M. Hardy, A.G. Flickinger, D.P. Riley, R.H. Weiss and U.S. Ryan, *J. Biol. Chem.*, 1994, **269**, 18 535.

39 D.P. Riley, P.J. Lennon, W.L. Neumann, R.H. Weiss, *J. Am. Chem. Soc.*, 1997, **119**, 6522.

40 J.M. Desper and S.H. Gellman, *J. Am. Chem. Soc.*, 1990, **112**, 6732.

41 H. Baker, P. Hambright, L. Wegner and L. Ross, *Inorg. Chem.*, 1973, **12**, 2200.

42 K.S. Kilgore, G.S. Freidrichs, C.R. Johnson, C.S. Schasteen, D.P. Riley, R.H. Weiss, U. Ryan and B.R. Lucchesi, *J. Mol. Cell. Cardiol.*, 1994, **26**, 995.

43 G.S. Freidrichs, K.S. Kilgore, L. Chi, U.S. Ryan and B.R. Lucchesi, *FASEB J.*, 1993, **7**, A425.

44 Y. Tamura, L. Chi, E.M. Driscoll Jr., P.T. Hoff, B.A. Freeman, K.P. Gallagher and B.R. Lucchesi, *Circulation Res.*, 1988, **63**, 944.

45 J.M. Downey, B. Omar, H. Ooiwa and J. McCord, *Free Radicals Res. Commun.*, 1991, **12–13**, 703.

46 S.C. Black, C.S. Schasteen, R.H. Weiss, D.P. Riley, E.M. Driscoll and B.R. Lucchesi, *J. Pharmacol. Exp. Therapeut.*, 1994, **270**, 1208.

47 C.M. Venturini, W.B. Sawyer, M.E. Smith, M.A. Palomo, E.G. McMahon, R.H. Weiss, D.P. Riley and C.S. Schasteen, in *The Biology of Nitric Oxide*, Vol. 3, ed. S. Moncada and E.A. Higgs. Portland Press, London, 1994.

48 T.P. Kasten, S. Settle, T.P. Misko, D.P. Riley, R.H. Weiss, M.G. Currie and G.A. Nickols, *Proc. Soc. Exp. Biol. Med.*, 1995, **208**, 170.

49 B. Halliwell, J.R. Hoult and D.R. Blake, *FASEB J.*, 1988, **2**, 2867.

50 M.B. Grisham, *Lancet*, 1994, **344**, 859.

51 R.H. Weiss, D.J. Fretland, D.A. Barron, U.S. Ryan, D.P. Riley, *J. Biol. Chem.*, 1996, **271**, 26 149.

52 J. Gaboury, R.C. Woodman, D.N. Granger, P. Reinhardt and P. Kubes, *Am. J. Physiol.*, 1993, **265**, H862.

53 P. Kubes, S. Kanwar, X.-F. Niu and J.P. Gaboury, *FASEB J.*, 1993, **7**, 1293.

54 X.-F. Niu, C.W. Smith and P. Kubes, *Circ. Res.*, 1994, **74**, 1133.

CHAPTER 6

Vanadium Compounds as Possible Insulin Modifiers

CHRIS ORVIG,[1] KATHERINE H. THOMPSON,[1]
MARGARET C. CAM[2] AND JOHN H. McNEILL[2]

[1]Department of Chemistry, University of British Columbia, 2036 Main Mall,
Vancouver BC, V6T 1Z1, Canada
[2]Faculty of Pharmaceutical Sciences, University of British Columbia, 2146
East Mall, Vancouver, BC, V6T 1Z3, Canada

1 Introduction

Vanadium has been used therapeutically on rare occasions since the turn of the century.[1] More recently, there has been a surge of interest in vanadium as an orally effective therapeutic agent due to its insulin-mimetic properties, which have been demonstrated *in vivo*.[2-4] Several recent volumes provide an overview of current research interests and findings in this rapidly expanding field of investigation.[5-7] An important aspect of this renewed interest is the potential for improved clinical efficacy of vanadium at very low concentrations by complexation with appropriate ligands.[8-12]

Insulin-mimetics are of interest principally as alternative or adjunct therapeutic agents in diabetes, a disease state characterized by an absolute or relative lack of insulin, and/or by insulin resistance. Insulin is a signaling hormone with numerous regulatory roles, including uptake of glucose, amino acids, and fatty acids for storage as, respectively, glycogen in muscle and liver, proteins in muscle, and triglycerides in adipose tissue.[13] Insulin also serves to counteract catabolic hormones, whose function is the mobilization of these molecular forms of stored energy. Insulin is not orally active.

In diabetes, glucose uptake into peripheral tissues such as skeletal muscle and fat is impaired, and glucose utilization in the energy-dependent processes within cells is abnormal.[14] The normal uptake and metabolism of glucose in non-diabetic individuals is initiated by a series of intracellular reactions known as the insulin signaling cascade.[15]

Insulin receptors (IRs) are membrane-spanning tyrosine-specific protein kinases;[16] early in the insulin signaling cascade, insulin, by binding on the extracellular side of cell membranes, activates the intracellular protein tyrosine phosphorylation of IRs. This step, and/or several protein kinases and phosphatases downstream, may be potential sites of action by vanadium (Figure 1).

In vitro studies of the insulin-mimetic actions of vanadium compounds are usually undertaken with phosphatases and kinases related to the insulin signaling cascade, or with enzymes relevant to glucose and lipid metabolism, in adipocytes or other tissues. For *in vivo* studies, a reproducible model of diabetes is obtained by administering streptozotocin (STZ) to rats at doses of $45-75\,mg\,kg^{-1}$ body weight. STZ-treated rats are insulinopenic, hyperphagic, and catabolic. The model does not completely parallel Type I diabetes in humans, in that STZ-diabetic rats can survive without administration of exogenous insulin; however, it is relatively simple, inexpensive, and reliable. Moderate to good diabetic control has been obtained in the STZ-diabetic rat, with several vanadium compounds, at oral doses of between 0.1 and $0.7\,mmol\,kg^{-1}\,day^{-1}$.[3,6,17-20] The dose of vanadium required to achieve good diabetic control varies with the initial diabetic state of the animal,[20] the particu-

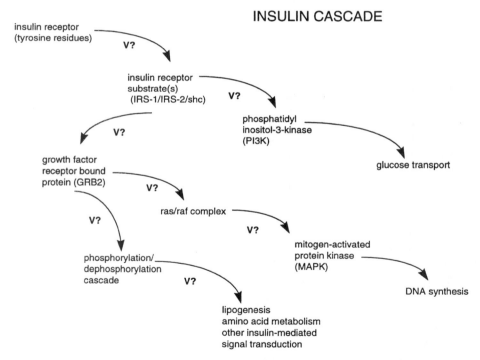

Figure 1 *Insulin signal transduction cascade (simplified). Intracellular kinases affected by tyrosine phosphorylation activation/deactivation under phosphotyrosine phosphatase (PTPase) regulation (vanadium-inhibitable) include (especially) IRS-1, IRS-2, shc, and MAPK. V? indicates possible sites of vanadium's mechanism of action. Cytosolic protein tyrosine kinase (CytPTK, not shown) stimulation by phosphatase inhibition is independent of the insulin cascade, but is also multi-step, and is particularly susceptible to vanadyl stimulation*

lar vanadium compound given,[21,22] addition of other trace elements[23] and as yet undetermined individual factors.[22,24] The BioBreeding (BB) Wistar rat is a spontaneous model of diabetes that closely resembles Type I diabetes. It is characterized by rapid onset of hyperglycemia at between 60 and 120 days of age and by loss of pancreatic insulin production. (Orally administered vanadium has effectively normalized blood glucose in this animal model of diabetes only when used in conjunction with exogenous insulin.[25])

Vanadium is a putative ultratrace element, although no biological role has been ascribed to it.[26] Reproducible and consistent deficiency symptoms have not been described; however, vanadium has been shown to be essential for growth and development of goats, rats, mice, and chicks.[27,28] It is normally present at very low concentrations ($< 10^{-8}$ M) in virtually all cells in plants and animals.[29] In man, the total body pool of vanadium is estimated to be 100 μg.[30,31] Most orally ingested vanadium is not absorbed, and therefore the predominant route of elimination is *via* the feces.[30] Vanadium is multivalent; however, the $+5$ and $+4$ oxidation states predominate in physiological environments (pH 3–7, aerobic solution, ambient temperature).[29,32–34] Vanadium(V) is generally present as orthovanadate, a mixture of $[HVO_4]^{2-}$ and $[H_2VO_4]^-$. Vanadium (IV) is found as the vanadyl ion, VO^{2+}. Vanadium in blood plasma exists in both oxidation states, balanced by oxygen tension and the presence of endogenous reducing agents such as ascorbate and catecholamines.[35] Physiological vanadium is largely protein bound (as are many other trace elements): to transferrin in plasma, to hemoglobin in erythrocytes, and to glutathione or other low-molecular-weight compounds intracellularly. The redox state of vanadium, whether $+4$ as vanadyl, or $+5$ as vanadate, depends on ligands, pH, and solute concentration. The majority of intracellular vanadium appears to be present as vanadyl, bound to small molecules or proteins, especially those containing thiol groups.[36,37] Vanadyl is known to be a less effective inhibitor of cellular phosphatases than vanadate;[38] however, there may be enough redox activity intracellularly to allow reversible formation of vanadate (and/or pervanadate) such that inhibition of tyrosine phosphatases may still be a crucial feature of vanadium's mechanism of insulin-like action.[39]

2 Characterization of Vanadium's Insulin-mimetic Effects

Characterization of the effects of inorganic vanadium on cell homogenates led to the first recognition of its insulin-mimetic effects. Sodium vanadate stimulated glucose uptake and glucose oxidation in rat adipocytes, stimulated glycogen synthesis in rat diaphragm and liver, and inhibited hepatic gluconeogenesis, usually at millimolar concentrations of added vanadate or vanadyl.[40,41] Lipid and glycolytic pathways were affected in specific tissues, mimicking insulin's effects, in most cases.

For instance, vanadate inhibited lipolysis and stimulated lipogenesis in adipocytes.[37,42] Glucose uptake and transport were enhanced by vanadate in other tissue types: mouse brain,[43] rat skeletal muscle,[44] and isolated villus cells.[45] Conversely, vanadate inhibited glucose transport in rat intestine when added to the perfusate of an everted gut sac preparation,[46] or when added as a supplement to the drinking water of diabetic and non-diabetic rats at concentrations up to 0.08 mM for 2–4 weeks.[47]

Stimulated translocation of the GLUT-4 glucose transporter (which is regulated by insulin) to the plasma membrane was demonstrated in rat adipocytes,[48] and increased glucose transporter expression in the presence of vanadate was shown *in vitro* in NIH 3T3 mouse fibroblasts[49] and *in vivo* in rat skeletal muscle.[50] Vanadate increased sensitivity to insulin in activating glucose transport in normal and insulin-resistant rat adipocytes[51] (but not human adipocytes[52]), *via* enhanced insulin binding.

Some insulin-like effects of vanadium are concentration-dependent. For example, in perfused rat liver, glucose output was inhibited by the addition of vanadate in the range of 0.5–1.0 μM, but was increased by concentrations greater that 25 μM.[53,54] In other experiments, *in vitro* and *in vivo* effects were found to be contradictory; for instance, when using a concentration range of 0.1–5.0 mM vanadate in rat hepatocytes, both non-insulin-like, glycogenolytic (glycogen breakdown) effects[55] and insulin-like, glycolytic effects[56] were demonstrated. Furthermore, these apparently opposing effects were also demonstrable in hepatocytes isolated from diabetic rats.[57,58] There have been conflicting reports, as well, on the effects of vanadium *in vitro* on pancreatic β cells: enhancement of glucose-stimulated insulin release in rat and mouse islets was observed at concentrations of 0.1–1.0 mM;[59,60] yet other investigators have shown inhibition at lower concentrations (1–50 μM).[61]

In vivo, vanadium effects on liver enzymes have tended to be more consistently insulin-like. Thus, in vanadate-treated diabetic rats having plasma concentrations of 17.5–22.0 μM, there was an increase in liver glycogen content, and enzymes involved in glycogen metabolism were affected in an insulin-like manner.[62] A series of *in vivo* studies has also consistently demonstrated that vanadate treatment of diabetic animals partially or completely restores liver and muscle enzyme activities involved in glycolysis and glycogenesis, such as glucokinase,[63] phosphoenolpyruvate carboxykinase,[18,64] pyruvate kinase,[65,66] and glycogen synthase,[67] and that these effects, at least on restored liver and muscle glycogen, are not secondary to a restoration of normal plasma glucose levels.[68] Vanadium treatment *in vivo* has been consistently and reproducibly associated with a decrease in plasma insulin in control animals,[2,18,24,69] resulting in an apparently enhanced peripheral insulin sensitivity[25] and a lowered insulin demand, although direct inhibitory effects cannot be completely ruled out. Differences between the reported effects *in vivo* and *in vitro* may be dependent on the actual concentrations of vanadium to which tissues are exposed.

3 Sites of Action of Vanadium

The sites of action of vanadium's insulin-mimetic effects would appear to be along the insulin cascade (Figure 1).[70] A hypothesis of insulin receptor (IR) tyrosine kinase stimulation and/or IR β-subunit phosphotyrosine phosphatase (PTPase) inhibition has received some experimental support.[71-75] Vanadate activated autophosphorylation of solubilized IRs, but did not phosphorylate serine or threonine residues of the receptor,[72,74,75] confirming its inhibitory action exclusively on tyrosine phosphatases.[71,73] Vanadate also activated tyrosine kinase in the IR β-subunit[76] and stimulated turnover of a phosphoryl group of pp15, a substrate of IR tyrosine kinase.[77] The activated tyrosine kinase activity induced by vanadate (10–200 μM) augmented insulin binding, enhanced insulin sensitivity, and prolonged insulin-stimulated lipolysis in rat adipocytes.[78,79]

Other studies have tended to rule out any involvement of the IR in vanadium's insulin-mimetic effects. *In vivo* studies showed that oral administration of vanadate produced profound insulin-like effects on glycogen metabolism without activating IR kinase.[80,81] Moreover, the activation of glucose transport by vanadate in rat adipocytes was shown to be unaffected by a loss of 60% of insulin receptors,[82] suggesting a post-receptor mechanism of action. Examination of tyrosine phosphorylation of the IR and other intracellular proteins in rat adipocytes revealed that vanadate had minimal effects on tyrosine phosphorylation at concentrations that produced a marked anti-lipolytic effect.[83] Also, oral vanadate treatment failed to change IR tyrosine kinase activity in STZ-diabetic rats, either isolated during the basal state or in response to a submaximal insulin clamp.[84,85] Vanadate may act at a post-receptor level through its inhibitory effects on several protein tyrosine phosphatases, or by a stimulatory effect on protein tyrosine kinases downstream from the IR. CytPTK, a cellular protein tyrosine kinase, was strongly activated by vanadate in rat adipocytes.[86] This effect was secondary to the inhibition of phosphotyrosine phosphatase, and may mediate vanadate-stimulated lipogenesis and glucose oxidation without affecting hexose uptake or inhibition of lipolysis.[87,88] Activation of CytPTK may result from the oxidation of vanadyl to vanadate by H_2O_2, thus implicating cellular H_2O_2 as a possible modulator of effects of the intracellular vanadyl insulin modification.[89]

Stimulation of insulin receptor kinase by peroxovanadates (complexes of vanadate and H_2O_2) presumably represents a mechanism of insulin mimesis different from that of vanadate or vanadyl.[11,90,91] Low concentrations (5–20 μM) of peroxovanadate, but not vanadate, potentiated insulin-stimulated glucose uptake in rat adipocytes, an effect that correlated with an increase in protein phosphotyrosine content.[11] Peroxovanadate inhibited lipolysis, stimulated protein synthesis and lipogenesis, and promoted autophosphorylation and activation of the IR tyrosine kinase, similarly to insulin.[92] Although peroxovanadate, unlike vanadate, increased tyrosine kinase activity in intact cells, neither compound, unlike insulin, had any effect on the kinase activity of partially purified adipocyte IR preparations.[90]

Vanadate, with added H_2O_2, markedly enhanced the IR kinase activity and increased the phosphorylation of at least four related proteins in rat hepatoma (Fao) cells, effects that were accompanied by inhibition of tyrosine phosphatase activity.[93] Recent evidence suggests the possibility of *in situ* formation of peroxovanadates, which may result in irreversible inhibition of some protein tyrosine phosphatases, as opposed to the readily reversible phosphatase inhibition seen with vanadates.[39] In addition, vanadate's substitution for phosphate within the enzyme structure may result in formation of a transition state analog of protein tyrosine phosphatase, thus changing the kinetics of tyrosine kinase activation, and effectively acting as a means of fine tuning the intracellular balance between phosphatase inhibition and kinase activation.[94,95]

The current preferred postulated mechanism of vanadium's *in vitro* effects is that vanadate stimulates specific protein-tyrosine phosphorylation by virtue of its inhibitory actions on appropriate PTPases.[39,96] Routine addition of vanadate to cell lysates, particularly for tyrosine kinase assays, is testimony of the ability of vanadate to preserve the phosphotyrosine content of cells.[97] The preponderance of evidence favors a post-receptor mechanism in stimulating glucose utilization, perhaps also involving a cytosolic (*i.e.* non-receptor) protein tyrosine kinase that is stimulated preferentially by vanadium and may be insulin independent.[86,87] Vanadyl is not a potent PTPase inhibitor.[38,97–99] Thus, whatever portion of intracellular vanadium is present as vanadyl is probably acting by some alternative mechanism.[35,96]

In the case of peroxovanadate[11,12] (which may be formed *in situ*),[39] vanadate plus superoxide radicals,[100] or vanadate with added hydrogen peroxide,[101,102] the mechanism appears to be increased autophosphorylation and activation of the insulin receptor, with consequent stimulation of glucose uptake. However, effects of intracellular vanadium on calcium influx[60,103] as well as intracellular and intravesicular pH modification[104–106] have not been ruled out as important factors in the mechanism of action of vanadium as an insulin-mimetic agent.

4   Animal Studies and Human Trials

The first definitive study to demonstrate *in vivo* glucose-lowering and anti-diabetic effects of vanadium was performed by McNeill and co-workers in 1985.[2] In this study, the addition of sodium orthovanadate ($0.39–0.54 \, \mathrm{mmol \, kg^{-1} \, day^{-1}}$) to the drinking water of STZ-diabetic rats over a period of six weeks dramatically lowered plasma glucose levels without improving depressed plasma insulin levels. Subsequent studies have shown that the glucose-lowering effect with orally administered vanadate or vanadyl solutions is persistent, relatively non-toxic, and accompanied by other desirable effects in plasma parameters, such as triglyceride and cholesterol lowering, and restoration of thyroxine levels to normal.[3,25,62,85,107] Although overall mean body weight in diabetic rats is not improved, the treatment consistently normalizes the elevated intake of food (hyperphagia) and fluids (polydipsia), both characteristic symptoms of STZ-diabetes, probably because of the normalization of metabolism, but also partly because of vanadium's anorexogenic effects.[108]

Decreases in plasma glucose levels and improved tolerance to an oral glucose tolerance test were recently reported to be similar regardless of the type of vanadium salt (vanadate or vanadyl) administered.[109] The apparent biphasic nature of the glucose-lowering response to chronic vanadium treatment[20,22,24] has led to the suggestion of a 'dual sensitivity' to oral vanadium treatment. Diabetic rats that were not rendered normoglycemic by low-dose treatment with vanadium (< 0.5 mmol kg^{-1} day^{-1}) nonetheless had significantly lower plasma triglyceride and cholesterol levels and improved glucose tolerance as compared to untreated diabetic animals, as well as a slight, but significant, improvement in pancreatic function as measured *in situ*.[22,110] It was subsequently found that the more severely diabetic rats (as determined by relatively higher blood glucose levels and lower plasma insulin levels before treatment) either could not be rendered normoglycemic or required higher concentrations of vanadyl to become so, which suggests that insulin and vanadyl work in a complementary manner *in vivo*.[20,111] The interrelationship of vanadium and insulin in the whole animal, suggested by *in vitro* studies, has also been confirmed by administration of insulin to vanadium-treated control and diabetic rats, wherein an increased sensitivity to insulin was seen.[25,112] In addition, treatment of the genetically diabetic BB rat reduces, although it does not completely eliminate, its exogenous insulin requirement.[25,113] That peripheral insulin resistance is also reversed by vanadium treatment has been shown by glucose clamp studies in which indwelling venous and arterial catheters permit infusion of tritiated glucose to maintain plasma glucose levels in the presence of high insulin levels (hyperinsulinemia). Using this method, hepatic glucose production and glucose utilization in response to submaximal or maximal insulin levels have been shown to be improved with vanadium treatment in STZ-diabetic rats[84,85] and in partially pancreatectomized rats.[114]

An intriguing discovery was that euglycemic and improved glucose tolerance, along with plasma, cardiac, and adipose tissue abnormalities were maintained for a long period (up to 30 weeks) after vanadyl treatment was withdrawn.[17,69,111] This may be attributed partly to an accumulation of vanadium in various tissue stores[115] or to the preservation or improvement of pancreatic β-cell insulin content in vanadium treated rats, leading to an indefinite period of near-normal glucose homeostasis in the fed state in these animals (unpublished observations).

The insulin-lowering effect of vanadium has particular relevance to Type II, or non-insulin dependent diabetes mellitus (NIDDM), in which hyperinsulinemia is a frequent concomitant and may be a contributing factor in the development of secondary complications of the disorder.[116] In animal models of Type II diabetes such as the *fa/fa* Zucker rat, vanadate treatment reduced food and fluid intake, reduced weight gain, lowered hyperinsulinemia towards normal levels, and improved glucose tolerance, whereas pair feeding only partially reversed these parameters.[117] Vanadate treatment improved insulin-mediated glucose utilization, without increasing glucose transporter levels in muscle.[118] In *ob/ob* mice, plasma glucose and insulin were reduced by 50% within one week of beginning treatment with vanadate.[119] Vanadate treatment preserved pancreatic insulin content, which in this animal model is normally exhausted as a result of hypersecretion. Glucose tolerance of genetically obese (*db/db*) diabetic mice was also improved by vanadate treatment.[120] In contrast to vanadium's effects on STZ-diabetic rats, no effect on plasma triglycerides was seen in models of Type II diabetes.

Reversal by vanadium treatment of a number of secondary complications relevant to diabetes has been reported. The development of cataracts, common in STZ-diabetic rats by 8 weeks after induction of diabetes, was eliminated with vanadyl treatment.[107,121,122] Sorbitol accumulation, which often precedes cataract development, was also reduced by oral vanadate treatment of diabetic rats.[123] Improved cardiac function in STZ-diabetic rats, as shown by working heart studies, resulted from orally administered vanadate[2] or vanadyl.[69] Abnormal adipose tissue function in diabetic rats, characterized by increased lipolytic rates, was corrected by vanadyl,[22,107] and the alterations in kidney morphology (swelling in the distal tubules) observed in untreated STZ diabetic rats were absent when they were treated with vanadyl.[124] Several plasma parameters that measure kidney function (blood urea nitrogen, creatinine) and liver function (AST), measures that are aberrant in the diabetic state, were reported to be normalized by vanadate or vanadyl.[3,69,107,122,124,125] Vanadyl treatment also resulted in near-normal organ/body weight ratios of lung, heart, liver, kidney, and the adrenal glands (as compared to the significantly elevated ratios observed in untreated diabetic rats).[122,125]

Recently, clinical trials of vanadium compounds have been initiated on human Type I (IDDM) and Type II (NIDDM) diabetic subjects.[126–128] Treatment with sodium orthovanadate ($125\,mg\,day^{-1}$) for 2 weeks resulted in significant increases in mean rates of glucose metabolism during a euglycemic clamp in two out of five subjects with IDDM and in five out of five subjects with NIDDM.[126] There were decreased insulin requirements in the IDDM subjects, and lowered serum cholesterol levels in all subjects. Treatment of NIDDM subjects with vanadyl sulphate ($100\,mg\,day^{-1}$) for 3 weeks caused an improved insulin sensitivity, a reduction in hepatic glucose production, and an increased rate of glucose disposal, which were sustained for 2 weeks after treatment was withdrawn.[127,128] In both of the studies, there were reported incidences of mild gastrointestinal intolerance.

Vanadyl therapy for 6 weeks at $50\,mg\,V\,day^{-1}$ resulted in improved insulin sensitivity in three of five human NIDDM subjects,[129] while basal hepatic glucose production was unchanged. Peak serum vanadium at this dose was $79.1 \pm 24.0\,ng\,ml^{-1}$. At $25\,mg\,V\,day^{-1}$ there was no change in glucose and lipid metabolic parameters and the peak serum vanadium was $15.7 \pm 3.7\,ng\,ml^{-1}$. There was no increase in thiobarbituric acid reactive substances (an indicator of *in vivo* lipid peroxidation) at these doses.

5 Toxicological Considerations

The type of toxicity most often associated with oral vanadium treatment is gastrointestinal, indicated by diarrhea and subsequent dehydration.[2,108] It has been suggested that vanadate is perhaps not as well tolerated as vanadyl;[130,131] however, differences may be slight.[109] The group of Domingo and colleagues has conducted a number of studies which showed an unfavorable toxicological profile of vanadium regardless of the salt administered, whether sodium meta-vanadate, sodium orthovanadate or vanadyl sulfate, with an increased incidence of mortality and the accumulation of vanadium in tissues.[131–134] In these studies, vanadium intake was usually, but not always associated with lowering of plasma glucose levels. Intraperitoneal administration of the iron chelator tiron with oral vanadate in STZ-diabetic rats lowered the accumulation of vanadium in several organs but did not diminish the anti-diabetic efficacy of vanadate.[135] However, a one-year toxicology study involving vanadyl sulfate at doses of 0.16–0.71 mmol kg^{-1} day^{-1} showed not only normalized plasma glucose and lipid levels in treated STZ-diabetic rats, but also no acceleration in the development of morphological abnormalities in a variety of organs (by histopathological tests) and no outstanding changes in hematological parameters.[122,125] Overall mortality was 19% in the vanadyl-treated diabetic rats, compared with 60% in the untreated diabetic rats. Tissue vanadium levels (at the end of one year) ranged from 6.5 to 15.1 μg g^{-1} in bone, 3.6 to 7.3 μg g^{-1} in kidney, and 0.2 to 0.4 μg g^{-1} in plasma.[121] Further reduction in such toxicological effects as decreased weight gain and reduced food intake, may be possible by using low-dose vanadium in mildly diabetic animals.[136]

6   Improved Tissue Uptake with Metal Chelation

An important advance in the use of vanadium compounds as insulin mimics has been the development of various ligands in order to improve substantially the absorption, tissue uptake and intracellular mobility of vanadium compounds, thereby reducing the dose required for optimal insulin mimesis. Selected coordination complexes of vanadium for which data (either *in vivo* or *in vitro*) have been published are shown in Figure 2. With the exception of the peroxovanadium(v) complexes, ligands have been chosen to chelate vanadyl; these ligands have generally been chosen to impart specific features to the resulting vanadium complexes: improved lipophilicity (vanadyl cysteine methyl ester, naglivan), improved oral absorption by passive diffusion (BMOV), potentiation of *in vitro* insulin-mimetic effect (the monoperoxo- and diperoxo-vanadates), or facilitation of transmembranal ion uptake (RL-252 and analogs). These are discussed below.

A series of bis(ligand)oxovanadium(IV) complexes were tested in STZ-diabetic rats;[8] ligands included salicylate, oxalate, malate, tartrate, and cysteine methyl ester. The vanadyl cysteine methyl ester complex, at a dose of 10 mg V kg^{-1} (0.2 mmol kg^{-1}) body weight by gavage was slightly more effective than other ligands in lowering plasma glucose levels within 24 hours of administration. There was no obvious toxicity at this dose; however, at ten times the glucose-lowering dose, all the test animals died of diarrhea within four days, suggesting the necessity for careful ligand design if the concept of chelated vanadyl is to prove worthwhile.

The water-insoluble vanadyl complex naglivan [bis(*N*-octylcysteineamide) oxo-vanadium(IV)] has been given to STZ-diabetic rats in a suspension of 3% acacia gum by oral gavage. Naglivan doses of 0.1–0.3 mmol kg^{-1} day^{-1} effective-

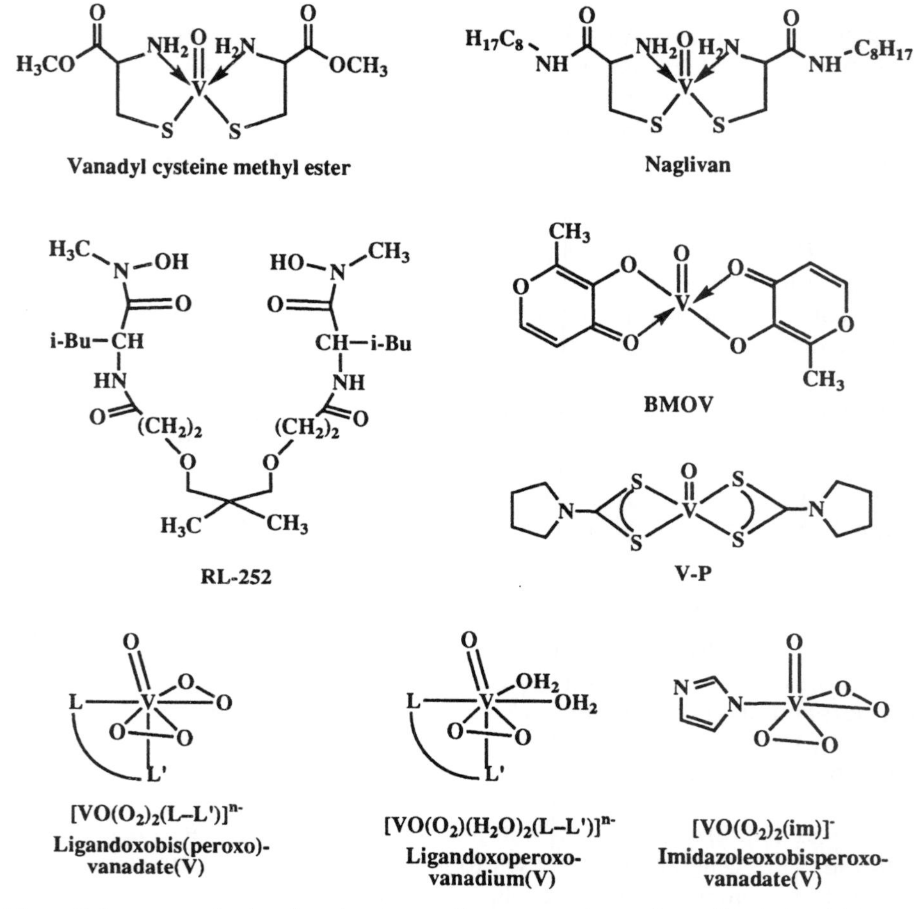

Vanadyl cysteine methyl ester

Naglivan

RL-252

BMOV

V-P

$[VO(O_2)_2(L-L')]^{n-}$
**Ligandoxobis(peroxo)-
vanadate(V)**

$[VO(O_2)(H_2O)_2(L-L')]^{n-}$
**Ligandoxoperoxo-
vanadium(V)**

$[VO(O_2)_2(im)]^-$
**Imidazoleoxobisperoxo-
vanadate(V)**

Figure 2 *Representative insulin-mimetic coordination complexes of vanadium(IV) and (V)*

ly lowered blood glucose levels to near normal, although the onset of action was significantly slower than with uncomplexed vanadate or vanadyl.[10,111] Naglivan treatment of both control and experimentally diabetic animals was not accompanied by weight loss or a reduction in food or fluid intake over an eight week period; however, neither was there diarrhea associated with naglivan treatment.

BMOV [bis(maltolato)oxovanadium(IV)] was designed as a vanadyl complex that would be water soluble, electrically neutral, of low molecular weight, and readily available for gastrointestinal absorption from drinking water (by passive diffusion).[9,137-139] Since the ligand, maltol, is an approved food additive, it was expected that the compound would also exhibit a positive safety profile. At a dose of $0.4 \, mmol \, kg^{-1} \, day^{-1}$, BMOV reduced blood glucose and lipid levels to near normal with no diarrhea and no mortality during a six-month test period.[21] The increased absorption of vanadium from BMOV was reflected in higher tissue vanadium concentrations, as compared with those from a similar treatment course with vanadyl sulfate.[110,140] In a number of acute experiments, BMOV was found to be 2–3 times as potent (in terms of glucose-lowering) as vanadyl sulphate, when administered by oral gavage or intraperitoneal injection.[141]

VP [bis(pyrrolidine-*N*-carbodithioato)oxovanadium(IV)] was initially tested as an *in vitro* insulin mimetic by inhibition of free fatty acid release from rat adipocytes.[142] It was administered orally to STZ diabetic rats at an initial dose (for 2 days) of $0.2 \, \text{mmol} \, \text{kg}^{-1} \, \text{day}^{-1}$ to achieve normoglycemia, followed by a maintenance dose of $0.1 \, \text{mmol} \, \text{kg}^{-1} \, \text{day}^{-1}$. Intraperitoneal administration of this compound proved to be more effective that oral treatment, but both achieved significant glucose-lowering.

A series of dihydroxamic acid chelators have been designed as hydrophobic carriers of vanadyl.[143] In an assay of lipogenic stimulation in rat adipocytes, RL-252 was maximally effective at molar ratios of 10:1 vanadyl sulfate:chelator, suggesting a shuttle mechanism of action. These compounds were electrically neutral, lipid-soluble, and optically chiral; they released the bound metal ion when treated with aqueous glutathione solutions.

In addition to organic ligands, vanadium has also been complexed with peroxide in order to potentiate its insulin-mimetic effect.[12,144] Complexation of vanadium(V) with hydrogen peroxide increased phosphorylation of the β-subunit,[90,92,145] stimulated lipogenesis, inhibited lipolysis, and promoted protein synthesis in rat adipocytes at micromolar concentrations of vanadium.[11,92] Peroxovanadate also increased the phosphorylation of several proteins, including those of the insulin receptor, when injected in the portal vein of rat livers.[146] Peroxovanadate lowered blood glucose in diabetic rats when administered intraperitoneally, but was ineffective when administered orally.[147,148] Because of the potent insulin-mimetic properties of these uncharacterised peroxovanadate mixtures, a variety of well-characterised monoperoxo- and diperoxo-vanadate(V) complexes have recently been prepared and tested *in vitro*, with very interesting results.[12,144] Twelve different peroxovanadium compounds of the $[\text{VO}(\text{O}_2)_2\text{L-L'}]^{n-}$ structure[12] inhibited phosphotyrosine phosphatase (PTPase), stimulated insulin receptor tyrosine kinase, and lowered plasma glucose when administered intraperitoneally. They also had a synergistic effect when co-administered with insulin.[12]

These potent insulin mimics activated insulin receptor tyrosine kinase and inhibited PTPase in rat liver endosomes at $5-80 \, \mu\text{M}$ vanadium concentrations, and lowered plasma glucose at doses in the $\text{mmol} \, \text{kg}^{-1}$ body weight range.[12] More recently, a six-coordinate bisperoxovanadium imidazole compound (see Figure 2) has been synthesized and shown to enhance insulin receptor autophosphorylation in human liver cell culture, as well as increase glucose transport in rat adipocytes, at concentrations ranging from $1 \, \mu\text{M}$ to $1 \, \text{mM}$.[149] The coordination of vanadium(V) to imidazole presents structural analogies to the coordination of vanadium to histidine residues in vanadium-containing haloperoxidases[150] and some phosphorylases.[151]

Combinations of vanadium and other trace elements have also been studied as a way to potentiate the action of vanadium. Pancreatectomized rats treated with a low concentration of oral sodium vanadate ($0.5 \, \text{g} \, \text{l}^{-1}$) and lithium carbonate showed improved glucose uptake and skeletal muscle glycogenic rate by glucose clamp.[152] Addition of magnesium and zinc to some of the vanadium- and lithium-treated rats produced a further slight, but significant, improvement in tissue glucose uptake. (Zn^{2+} has also been found to mimic several actions of insulin, *in vitro* and *in vivo*, by a mechanism unrelated to insulin.[153]) In separate experiments, diabetic rats co-administered sodium vanadate and lithium in their drinking water, were rendered normoglycemic within 4 days, and had partially restored liver and kidney superoxide dismutase activity after 16 days of treatment.[23]

7 Summary

The potential of vanadium compounds to mimic insulin when administered orally has stimulated renewed interest in the pharmacological effects of these unique and enigmatic compounds. Both organically-chelated and inorganic vanadium compounds have been shown to lower blood glucose and ameliorate other diabetic symptoms in a variety of animal models of both insulin-dependent and non-insulin-dependent diabetes. Diabetes is characterized by insulin deficiency or insulin resistance. Because insulin cannot be absorbed intact following oral administration and must be administered parenterally, available therapies are cumbersome at best. Determining a therapeutic dosing regimen without significant toxicity for orally administered vanadium compounds would present a significant advance over currently available treatments.

References

1 B.M. Lyonnet and E. Martz Martin, *La Presse Médicale*, 1899, 7, 191.

2 C.E. Heyliger, A.G. Tahiliani and J.H. McNeill, *Science*, 1985, **227**, 1474.

3 J. Meyerovitch, Z. Farfel, J. Sack and Y. Shechter, *J. Biol. Chem.*, 1987, **262**, 6658.

4 S.M. Brichard, J. Lederer and J.-C. Henquin, *Diabete Metabolisme (Paris)*, 1991, **17**, 435.

5 H. Sigel and A. Sigel, *Metal Ions in Biological Systems*, Marcel Dekker, Inc., New York, Vol. 31, *Vanadium and Its Role in Life*.

6 N.D. Chasteen, *Vanadium in Biological Systems: Physiology and Biology*, Kluwer Academic Publishers, Dordrecht, 1990.

7 A.K. Srivastava and J.-L. Chiasson, *Vanadium Compounds: Biochemical and Therapeutic Applications, Mol. Cell Biochem.*, **153**, Kluwer Academic Publishers, Dordrecht, 1995.

8 H. Sakurai, K. Tsuchiya, M. Nukatsuka, J. Kawada, S. Ishikawa, H. Yoshida and M. Komatsu, *J. Clin. Biochem. Nutrit.*, 1990, **8**, 193.

9 J.H. McNeill, V.G. Yuen, H.R. Hoveyda and C. Orvig, *J. Med. Chem.*, 1992, **35**, 1489.

10 M.C. Cam, G.H. Cros, J.-J. Serrano, R. Lazaro and J.H. McNeill, *Diab. Res. Clin. Prac.*, 1993, **20**, 111.

11 A. Shisheva and Y. Shechter, *Endocrinology*, 1993, **133**, 1562.

12 B.I. Posner, R. Faure, J.W. Burgess, A.P. Bevan, D. Lachance, G. Zhang-Sun, I.G. Fantus, J.B. Ng, D.A. Hall, B. Soo Lum and A. Shaver, *J. Biol. Chem.*, 1994, **269**, 4596.

13 M.P. Czech, *Ann. Rev. Biochem.*, 1977, **46**, 359.

14 G.E. Lienhard, J.W. Slot, D.E. James and M.M. Mueckler, *Sci. Am.*, 1992, **267**, 86.

15 C.R. Kahn and M.F. White, *J. Clin. Invest.*, 1988, **82**, 1151.

16 M.F. White, S.E. Shaleson, H. Keutmann and C.R. Kahn, *J. Biol. Chem.*, 1988, **263**, 2969.

17 R.A. Pederson, S. Ramanadham, A.M.M. Buchan and J.H. McNeill, *Diabetes*, 1989, **38**, 1390.

18 O. Blondel, J. Simon, B. Chevalier and B. Portha, *Am. J. Physiol.*, 1990, **258**, E459.

19 N. Sekar, A. Kanthasamy, S. Williams, S. Subramanian and S. Govinsasamy, *Pharmacol. Res.*, 1990, **22**, 207.

20 K.H. Thompson, J. Leichter and J.H. McNeill, *Biochem. Biophys. Res. Commun.*, 1993, **197**, 1549.

21 V.G. Yuen, C. Orvig and J.H. McNeill, *Can. J. Physiol. Pharmacol.*, 1993, **71**, 263.

22 M.C. Cam, R.A. Pederson, R.W. Brownsey and J.H. McNeill, *Diabetologia*, 1993, **36**, 218.

23 P. Srivastava, A.K. Saxena, R.K. Kale and N.Z. Baquer, *Res. Commun. Chem. Path. Pharmacol.*, 1993, **80**, 283.

24 M. Bendayan and D. Gingras, *Diabetologia*, 1989, **32**, 561.

25 S. Ramanadham, G.H. Cros, J.J. Mongold, J.J. Serrano and J.H. McNeill, *Can. J. Physiol. Pharmacol.*, 1990, **68**, 486.

26 I.G. Macara, *Trends Biochem. Sci.*, 1980, **5**, 92.

27 N.D. Chasteen, *Struct. Bonding (Berlin)*, 1983, **53**, 105.

28 B.R. Nechay, *Ann. Rev. Pharmacol. Toxicol.*, 1984, **24**, 501.

29 D. Rehder, *Angew. Chem., Int. Ed. Engl.*, 1991, **103**, 148.

30 A.R. Byrne and L. Kosta, *Sci. Total Envir.*, 1978, **10**, 17.

31 R. Cornelis, J. Versieck, L. Mees, J. Hoste and F. Barbier, *Biol. Trace Element Res.*, 1981, **3**, 257.

32 R. Wever and K. Kustin, *Adv. Inorg. Chem.*, 1990, **35**, 81.

33 F.H. Nielsen and E.O. Uthus, in *Vanadium in Biological Systems*, ed. N.D. Chasteen, Kluwer, Dordrecht, 1990, p. 51.

34 A.R. Butler and C.J. Carrano, *Coord. Chem. Rev.*, 1991, **109**, 61.

35 M. Bruech, M.E. Quintanilla, W. Legrum, J. Koch, K.J. Netter and G.F. Fuhrmann, *Toxicology*, 1984, **31**, 283.

36 N.D. Chasteen, J.K. Grady and C.E. Holloway, *Inorg. Chem.*, 1986, **25**, 2754.

37 H. Degani, M. Gochin, S.J.D. Karlish and Y. Shechter, *Biochemistry*, 1981, **20**, 5795.

38 L.C. Cantley, Jr. and P. Aisen, *J. Biol. Chem.*, 1979, **254**, 1781.

39 G. Huyer, S. Liu, J. Kelly, J. Moffat, P. Payette, B. Kennedy, G. Tsaprailis, M.J. Gresser and C. Ramachandran, *J. Biol. Chem.*, 1997, **272**, 843.

40 Y. Shechter and S.J.D. Karlish, *Nature (London)*, 1980, **284**, 556.

41 G.R. Dubyak and A. Kleinzeller, *J. Biol. Chem.*, 1980, **255**, 5306.

42 W.C. Duckworth, S.S. Solomon, J. Liepnieks, F.G. Hamel, S. Hand and D.E. Peavy, *Endocrinology*, 1988, **122**, 2285.

43 S. Amir, J. Meyerovitch and Y. Shechter, *Brain Res.*, 1987, **419**, 392.

44 N. Okumura and T. Shimazu, *J. Biochem. (Tokyo)*, 1992, **112**, 107.

45 J.J. Hajjar, M.P. Dobish and T.K. Tomicic, *Proc. Soc. Exp. Biol. Med.*, 1989, **190**, 35.

46 G.L. Kellett and E.D. Barker, *Biochim. Biophys. Acta*, 1989, **979**, 311.

47 K.L. Madsen, V.M. Porter and R.N. Fedorak, *Diabetes*, 1993, **42**, 1126.

48 M.R. Paquet, R.J. Romanek and R.J. Sargeant, *Mol. Cell. Biochem.*, 1992, **109**, 149.

49 K.G. Mountjoy and J.S. Flier, *Endocrinology*, 1990, **127**, 2025.

50 H.V. Strout, P.P. Vicario, C. Biswis, R. Saperstein, E.J. Brady, P.F. Pilch and J. Berger, *Endocrinology*, 1990, **126**, 2728.

51 J.W. Eriksson, P. Lonnroth and U. Smith, *Diabetologia*, 1992, **35**, 510.
52 P. Lonroth, J.W. Eriksson, B.I. Posner and U. Smith, *Diabetologia*, 1993, **36**, 113.
53 R. Bruck, H. Prigozin, Z. Krepel, P. Rotenberg, Y. Shechter and S. Bar-Meir, *Hepatology*, 1991, **14**, 540.
54 M. Roden, K. Liener, C. Fürnsinn, M. Prskavec, P. Nowotny, I. Steffan, H. Vierhapper and W. Waldhäusl, *Diabetologia*, 1993, **36**, 602.
55 F. Bosch, J. Arino, A.M. Gomez-Foix and J.J. Giunovart, *J. Biol. Chem.*, 1987, **262**, 218.
56 A.M. Gomez-Foix, J.E. Rodriguez-Gil, C. Fillat, J.J. Guinovart and F. Bosch, *Biochem. J.*, 1988, **255**, 507.
57 J.E. Rodriguez-Gil, A.M. Gomez-Foix, J. Arino, J.J. Guinovart and F. Bosch, *Diabetes*, 1989, **38**, 793.
58 J.E. Rodriguez-Gil, A.M. Gomez-Foix, C. Fillat, F. Bosch and J.J. Guinovart, *Diabetes*, 1991, **40**, 1355.
59 J.A. Fagin, K. Ikejiri and S.R. Levin, *Diabetes*, 1987, **36**, 1448.
60 A.Q. Zhang, Z.Y. Gao, P. Gilon, M. Nenquin, G. Drews and J.C. Henquin, *J. Biol. Chem.*, 1991, **266**, 21 649.
61 C. Voss, I. Herrmann, K. Hartmann and H. Zuhlke, *Exp. Clin. Endocrinol.*, 1992, **99**, 159.
62 S. Pugazhenthi and R.L. Khandelwal, *Diabetes*, 1990, **39**, 821.
63 J. Gil, M. Miralpeix, J. Carreras and R. Bartrons, *J. Biol. Chem.*, 1988, **263**, 1868.
64 A. Valera, J.E. Rodriguez-Gil and F. Bosch, *J. Clin. Invest.*, 1993, **92**, 4.
65 A.K. Saxena, P. Srivastava and N.Z. Baquer, *Eur. J. Pharmacol.*, 1992, **216**, 123.
66 S.M. Brichard, B. Desbuquois and J. Girard, *Mol. Cell. Endocrinol.*, 1993, **91**, 91.
67 M. Bollen, M. Miralpeix, F. Ventura, B. Toth, R. Bartrons and W. Stalmans, *Biochem. J.*, 1990, **267**, 269.
68 L. Rossetti and M.R. Laughlin, *J. Clin. Invest.*, 1989, **84**, 892.
69 S. Ramanadham, J.J. Mongold, R.W. Brownsey, G.H. Cros and J.H. McNeill, *Am. J. Physiol.*, 1989, **257**, H904.
70 E. Tsiani and I.G. Fantus, *Trends Endocrinol. Metab.*, 1997, **8**, S1.
71 G. Swarup, K.V.J. Speeg, S. Cohen and D.I. Garbers, *J. Biol. Chem.*, 1982, **257**, 7298.
72 S. Tamura, T.A. Brown, R.E. Dubler and J. Larner, *Biochem. Biophys. Res. Commun.*, 1983, **113**, 80.
73 A.S. Tracey and M.J. Gresser, *Proc. Natl. Acad. Sci. USA*, 1986, **83**, 609.
74 A. Ueno, N. Arakaki, Y. Takeda and H. Fujio, *Biochem. Biophys. Res. Commun.*, 1987, **144**, 11.
75 R. Gherzi, C. Caratti, G. Andraghetti, S. Bertolini, A. Montemurro, G. Sesti and R. Cordera, *Biochem. Biophys. Res. Commun.*, 1988, **152**, 1474.
76 D.M. Smith and G.J. Sale, *Biochem. J.*, 1988, **256**, 903.
77 M. Bernier, D.M. Laird and M.D. Lane, *J. Biol. Chem.*, 1988, **263**, 13 626.
78 I.G. Fantus, F. Ahmad and G. Deragon, *Endocrinology*, 1990, **127**, 2716.
79 I.G. Fantus, F. Ahmad and G. Deragon, *Diabetes*, 1994, **43**, 375.
80 T.K. Jackson, A.I. Salhanick, J.D. Sparks, C.E. Sparks, M. Bolognino and J.M. Amatruda, *Diabetes*, 1988, **37**, 1234.
81 H.V. Strout, P.P. Vicario, R. Saperstein and E.E. Slater, *Endocrinology*, 1989, **124**, 1918.
82 A. Green, *Biochem. J.*, 1986, **238**, 663.
83 R.A. Mooney, K.L. Bordwell, S. Luhowskyj and J.E. Casnelle, *Endocrinology*, 1989, **124**, 422.
84 O. Blondel, D. Bailbe and B. Portha, *Diabetologia*, 1989, **32**, 185.
85 N. Venkatesan, A. Avidan and M.B. Davidson, *Diabetes*, 1991, **40**, 492.
86 A. Shisheva and Y. Shechter, *FEBS Lett.*, 1992, **300**, 93.

87 A. Shisheva and Y. Shechter, *J. Biol. Chem.*, 1993, **268**, 6463.

88 J. Li, G. Elberg, D.C. Crans and Y. Shechter, *Biochemistry*, 1996, **35**, 8314.

89 G. Elberg, J. Li and Y. Shechter, *J. Biol. Chem.*, 1994, **269**, 9521.

90 S. Kadota, G. Fantus, G. Deragon, H.J. Guyda, B. Hersh and B.I. Posner, *Biochem. Biophys. Res. Commun.*, 1987, **147**, 259.

91 D. Heffetz, W.J. Rutter and Y. Zick, *Biochem. J.*, 1992, **288**, 631.

92 I.G. Fantus, S. Kadota, G. Deragon, B. Foster and B.I. Posner, *Biochemistry*, 1989, **28**, 8864.

93 D. Heffetz, I. Bushkin, R. Dror and Y. Zick, *J. Biol. Chem.*, 1990, **265**, 2896.

94 J.M. Denu, D.L. Lohse, J. Vijayalakshmi, M.A. Saper and J.E. Dixon, *Proc. Natl. Acad. Sci. USA*, 1996, **93**, 2493.

95 M. Zhang, M. Zhou, R.L. Van Etten and C.V. Stauffacher, *Biochemistry*, 1997, **36**, 15.

96 G. Elberg, Z. He, J. Li, N. Sekar and Y. Shechter, *Diabetes*, 1997, **46**, 1684.

97 J.A. Gordon, *Methods Enzmol.*, 1991, **201**, 477.

98 I.G. Macara, K. Kustin and L.C. Cantley, Jr., *Biochim. Biophys. Acta*, 1980, **629**, 95.

99 A. Barberà, J.E. Rodriguez-Gil and J.J. Guinovart, *J. Biol. Chem.*, 1994, **269**, 20 047.

100 S. Trudel, M.R. Paquet and S. Grinstein, *Biochem. J.*, 1991, **276**, 611.

101 Y. Zick and R. Sagi-Eisenberg, *Biochemistry*, 1990, **29**, 10 240.

102 D. Hecht and Y. Zick, *Biochem. Biophys. Res. Commun.*, 1992, **188**, 773.

103 D.M. Delfert and J.M. McDonald, *Arch. Biochem. Biophys.*, 1985, **241**, 665.

104 D. Cassel, Y.-X. Zhuang and L. Glaser, *Biochem. Biophys. Res. Commun.*, 1984, **118**, 675.

105 Y. Shechter, J. Meyerovitch, Z. Farfel, J. Sack, R. Bruck, S. Bar-Meir, S. Amir, H. Degani and S.J.D. Karlish, in *Vanadium in Biological Systems*, ed. N.D. Chasteen, Kluwer, Dordrecht, 1990, p. 129.

106 P. Munoz, A. Guma, M. Camps, M. Furriols, X. Testar, M. Palacin and A. Zorzano, *J. Biol. Chem.*, 1992, **267**, 10 381.

107 S. Ramanadham, R.W. Brownsey, G.H. Cros, J.J. Mongold and J.H. McNeill, *Metabolism*, 1989, **38**, 1022.

108 J. Meyerovitch, Y. Shechter and S. Amir, *Physiol. Behavior*, 1989, **45**, 1113.

109 D. J. Becker, L.-N. Ongemba and J.-C. Henquin, *Eur. J. Pharmacol.*, 1994, **260**, 169.

110 V.G. Yuen, C. Orvig, K.H. Thompson and J.H. McNeill, *Can. J. Physiol. Pharmacol.*, 1993, **71**, 270.

111 M.C. Cam, J. Faun and J.H. McNeill, *Metab. Clin. Exp.*, 1995, **44**, 332.

112 R.A.J. Challiss, B. Leighton, F.J. Lozeman, L. Budohoski and E.A. Newsholme, *Biochem. Pharmacol.*, 1987, **36**, 357.

113 M.L. Battell, V.G. Yuen and J.H. McNeill, *Pharmacol. Commun.*, 1992, **1**, 291.

114 R. Cordera, F. Andraghetti, R.A. DeFronzo and L. Rossetti, *Endocrinology*, 1990, **126**, 2177.

115 R.D.R. Parker and R.P. Sharma, *J. Environ. Pathol. Toxicol.*, 1978, **2**, 235.

116 G.M. Reaven, *Annu. Rev. Med.*, 1993, **44**, 121.

117 S.M. Brichard, A.M. Pottier and J.C. Henquin, *Endocrinology*, 1989, **125**, 2510.

118 S.M. Brichard, L.N. Ongemba and J.-C. Henquin, *Diabetologia*, 1992, **35**, 522.

119 S.M. Brichard, C.J. Bailey and J.-C. Henquin, *Diabetes*, 1990, **39**, 1326.

120 S. Pugazhenthi, J.F. Angel and R.L. Khandelwal, *Metabolism*, 1991, **40**, 941.

121 K.H. Thompson and J.H. McNeill, *Res. Commun. Chem. Path. Pharmacol.*, 1993, **80**, 187.

122 S. Dai, K.H. Thompson and J.H. McNeill, *Pharmacol. Toxicol.*, 1994, **74**, 101.

123 J.W. Lohr, M.I. Bennett, M.A. Pochal, J. McReynolds, M. Acara and G.R. Willsky, *Res. Commun. Chem. Pathol. Pharmacol.*, 1991, **72**, 191.

108 *Vanadium Compounds as Possible Insulin Modifiers*

124 J.J. Mongold, G.H. Cros, L. Vian, A. Tep, S. Ramanadham, G. Siou, J. Diaz, J.H. McNeill and J.J. Serrano, *Pharmacol. Toxicol.*, 1990, **67**, 192.
125 S. Dai and J.H. McNeill, *Pharmacol. Toxicol.*, 1994, **74**, 110.
126 A.B. Goldfine, D.C. Siminson, F. Folli, M.-E. Patti and C.R. Kahn, *J. Clin. Endocrinol. Metab.*, 1995, **80**, 3311.
127 N. Cohen, M. Halberstam, P. Shlimovich, C.J. Chang, H. Shamoon and L. Rossetti, *J. Clin Invest.*, 1995, **95**, 2501.
128 M. Halberstam, N. Cohen, P. Shlimovich, L. Rosetti, and H. Shamoon, *Diabetes*, 1996, **45**, 659.
129 A.B. Goldfine, personal communication.
130 J.M. Llobet and J.L. Domingo, *Toxicol. Lett.*, 1984, **23**, 227.
131 J.L. Domingo, M. Gomez, J.M. Llobet, J. Corbella and C.L. Keen, *Pharmacol. Toxicol.*, 1991, **68**, 249.
132 J.L. Domingo, A. Ortega, J.M. Llobet and C.L. Keen, *Trace Elements Med.*, 1991, **8**, 181.
133 J.L. Domingo, M. Gomez, J.M. Llobet, J. Corbella and C.L. Keen, *Toxicology*, 1991, **66**, 279.
134 J.L. Domingo, D.J. Sanchez, M. Gomez, J.M. Llobet and J. Corbella, *Res. Commun. Chem. Path. Pharmacol.*, 1992, **75**, 369.
135 J. L. Domingo, D. J. Sanchez, M. Gomez, J. M. Llobet and J. Corbella, *Vet. Human Toxicol.*, 1993, **35**, 495.
136 J.C. Henquin, F. Carton, L.N. Ongemba and D.J. Becker, *J. Endocrin.*, 1994, **142**, 555.
137 P. Caravan, L. Gelmini, N. Glover, F.G. Herring, H. Li, J.H. McNeill, S.J. Rettig, I.A. Setyawati, E. Shuter, Y. Sun, A.S. Tracey, V.G. Yuen and C. Orvig, *J. Am. Chem. Soc.*, 1995, **117**, 12759.
138 Y. Sun, B.R. James, S.J. Rettig and C. Orvig, *Inorg. Chem.*, 1996, **35**, 1667.
139 G.R. Hanson, Y. Sun and C. Orvig, *Inorg. Chem.*, 1996, **35**, 6507.
140 I.A. Setyawati, K.H. Thompson, V.G. Yuen, Y. Sun, M. Battell, D.M. Lyster, C. Vo, T.J. Ruth, S. Zeisler, J.H. McNeill and C. Orvig, *J. Appl. Physiol.*, 1998, **84**, 569.
141 V.G. Yuen, C. Orvig and J.H. McNeill, *Can. J. Physiol. Pharmacol.*, 1995, **73**, 55.
142 H. Watanabe, M. Nakai, K. Komazawa and H. Sakurai, *J. Med. Chem.*, 1994, **37**, 876.
143 Y. Shechter, A. Shisheva, R. Lazar, J. Libman and A. Shanzer, *Biochemistry*, 1992, **31**, 2063.
144 A. Shaver, J.B. Ng, D.A. Hall, B. Soo Lum and B.I. Posner, *Inorg. Chem.*, 1993, **32**, 3109.
145 S. Kadota, I.G. Fantus, G. Deragon, H.J. Guyda and B.I. Posner, *J. Biol. Chem.*, 1987, **262**, 8252.
146 Y.R. Hadari, B. Geiger, O. Nadiv, I. Sabanay, C.T. Roberts, D. Leroith and Y. Zick, *Mol. Cell. Endocrinology*, 1993, **97**, 9.
147 A. Shisheva, O. Ilonomov and Y. Shechter, *Endocrinology*, 1994, **134**, 507.
148 A.P. Bevan, P.G. Drake, J.-F. Yale, A. Shaver and B.I. Posner, *Mol. Cell. Biochem.*, 1995, **153**, 49.
149 D.C. Crans, A.D. Keramidas, H. Hoover-Litty, O.P. Anderson, M.M. Miller, L.M. Lemoine, S. Pleasic-Williams, M. Vandenberg, A.J. Rossomando and L.J. Sweet, *J. Am. Chem. Soc.*, 1997, **119**, 5447.
150 A. Messerschmidt and R. Wever, *Proc. Natl. Acad. Sci. USA*, 1996, **93**, 392.
151 Y. Lindqvist, G. Schneider and P. Vihko, *Eur. J. Biochem.*, 1994, **221**, 139.
152 L. Rossetti, A. Giaccari, E. Klein-Robbenhaar and L.R. Vogel, *Diabetes*, 1990, **39**, 1243.
153 A. Shisheva, D. Gefel and Y. Shechter, *Diabetes*, 1992, **41**, 982.

CHAPTER 7

Cisplatin-based Anticancer Agents

LLOYD R. KELLAND

CRC Centre for Cancer Therapeutics, The Institute of Cancer Research,
15 Cotswold Road, Belmont, Sutton, Surrey SM2 5NG, UK

1 Introduction

The anticancer drugs based on platinum: cisplatin [CDDP, cis-diamminedichloroplatinum(II)], and its analogue carboplatin [CBDCA, cis-diammine(1,1-cyclobutanedicarboxylato)platinum(II)] are among those most frequently used today. Interestingly, the chemical identity of cisplatin was first established in the mid-19th century as Peyrone's chloride, and, but for the fortuitous and well-documented observations of Barnett Rosenberg while performing experiments to investigate the effects of electric fields on bacteria, its antitumour properties might have remained unknown.[1]

This chapter describes the clinical properties of cisplatin, and of its analogues that have undergone extensive clinical trials (particularly carboplatin), platinum chemistry relating to mechanism of action, the mechanisms of tumour resistance to cisplatin and their circumvention.

2 Clinical Properties

Cisplatin was introduced into clinical practice in 1971 (only some five years after the initial discovery of its cell-killing properties), and the less toxic analogue, carboplatin, in 1981. To date, carboplatin is the only platinum analogue to have received worldwide registration. Comparative clinical properties of cisplatin and carboplatin are summarised in Table 1.

Undoubtedly the most dramatic impact of cisplatin has been observed in men presenting with testicular cancer. Before 1975, the cure of such patients, who are predominately young adults, was rare. Following the introduction of cisplatin into a regimen also containing vinblastine and bleomycin (PVB regime),[2] around 85% of these patients now are essentially cured of the disease.

Initial studies at the Royal Marsden Hospital (London) by Wiltshaw and

Table 1 *Comparative clinical properties of cisplatin and carboplatin*

	Cisplatin	Carboplatin
Typical dose/schedule		
	100 mg m^{-2}; i.v. q 3–4 weeks or 20 mg m^{-2} daily for 5 days, q 3–4 weeks	400 mg m^{-2}; i.v. q 3–4 weeks
Major toxicities		
	Nephrotoxicity Severe nausea and vomiting Neurotoxicity (peripheral neuropathy) Ototoxicity (tinnitus/hearing loss)	Myelosuppression (mainly thrombocytopaenia)
Pharmacokinetics[22]		
(A) Total platinum		
$t_{1/2}\,\alpha$	20–40 min	1–3 h
$t_{1/2}\,\beta$	44–190 h	6.7– > 24 h
(B) Free (ultrafilterable) platinum		
$t_{1/2}\,\alpha$	22–78 min	87 min
$t_{1/2}\,\beta$	not seen	354 min
24 h Urinary platinum excretion (% of dose)[22]		
	16–35	65

colleagues established that cisplatin also conferred promising activity against ovarian cancer.[3] Combination regimens including cisplatin (typically with cyclophosphamide) produce clinical complete remissions in approximately 50% of patients with advanced disease. Accrued long-term survival data from the Netherlands for women presenting with advanced ovarian cancer have shown that combination chemotherapy with cisplatin can improve survival rates at 5 and 10 years by more than 10% as compared with the best available treatment of the pre-cisplatin era.[4] Cisplatin also confers a major palliative effect in patients with small-cell lung carcinoma, bladder carcinoma, head and neck carcinoma, or cervical carcinoma.[5]

Cisplatin *versus* Carboplatin

Numerous clinical trials have demonstrated that carboplatin is substantially less toxic (especially in terms of nephrotoxicity and gastrointestinal effects; see Table 1) than is cisplatin. Overview analyses of randomised studies comparing the activity of cisplatin *versus* that of carboplatin (mainly in ovarian and testicular cancers) have concluded that the two are broadly comparable in terms of response rates and disease-free intervals.[6–8] However, it also appears that the two drugs are effective against the same population of tumours and thus share cross-resistance with one another.[9,10] Notably, however, secondary responses to either cisplatin or carboplatin may occur in patients with ovarian cancer who have previously responded to the drugs; response rates increase with increasing progression-free intervals.[11]

While some have advocated replacing cisplatin with the less toxic carboplatin as first-line treatment for all patients with ovarian tumours,[12] this issue remains contentious, both with respect to cost and because many of the randomised studies of cisplatin *versus* carboplatin are still, in the context of clinical trials, in relatively early stages without long-term follow up.

Iproplatin

Iproplatin [CHIP, *cis*-dichloro-*trans*-dihydroxo-bis(isopropylamine) platinum(IV); see Figure 1 for structure], like carboplatin, was selected for clinical evaluation because of its favourable efficacy profile in preclinical studies, *i.e.* less nephrotoxicity but antitumour activity comparable to that of cisplatin.[13] Iproplatin was the first quadrivalent platinum(IV) complex possessing an octahedral configuration, rather than the square-planar configuration of cisplatin and

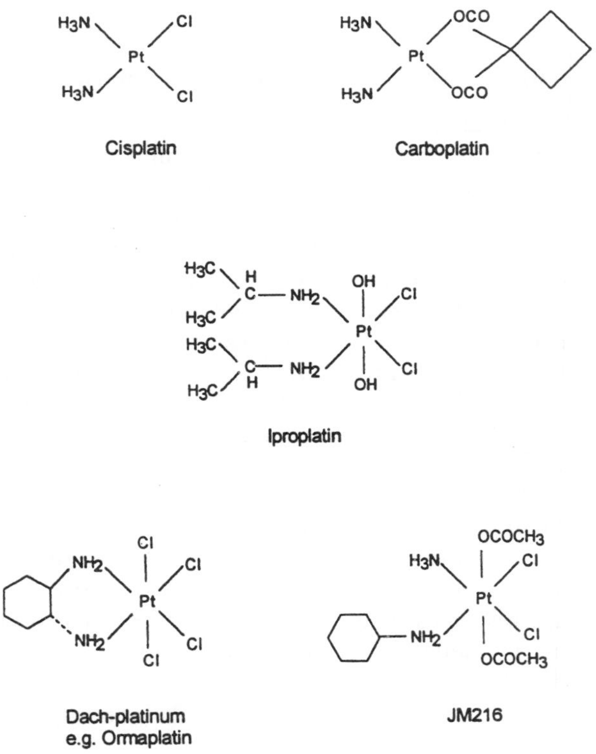

Figure 1 *Structures of platinum drugs that have undergone extensive clinical trials (cisplatin, carboplatin and iproplatin) or are new drugs currently in early clinical trials*

carboplatin, that entered clinical trials. To date, it is the only platinum drug besides cisplatin and carboplatin to have undergone extensive (including phase III) clinical trials.

The three major conclusions from these clinical trials are:

(a) Iproplatin has a toxicity profile similar to that of carboplatin, with myelosuppression (mainly thrombocytopaenia), being dose-limiting.[14]
(b) Iproplatin does confer responses in 'platinum-sensitive' tumour types such as ovarian cancer, but shares cross-resistance with cisplatin and carboplatin.[14,15]
(c) Most importantly, in randomised trials with carboplatin, iproplatin was more toxic (especially to the gastrointestinal tract) and significantly less active.[16,17] For example, in a randomised trial of 120 patients with advanced ovarian cancer, median survival values were, respectively, 114 weeks and 68 weeks ($p = 0.008$) for carboplatin ($400 \, mg \, m^{-2}$ every 4 weeks) and iproplatin ($300 \, mg \, m^{-2}$ every 4 weeks).[16]

In conclusion, iproplatin appears to be less active and more toxic than carboplatin. Its future clinical role is limited.

Determination of Platinum Drug Levels and Pharmacokinetics

Platinum levels in cells are generally measured by flameless atomic absorption spectrometry (FAAS), a technique that uses high temperatures to atomise molecules, and that specifically measures the absorbance of platinum at a wavelength of 265.9 nm. The measurement of platinum following therapeutic doses of drug (levels down to about $5 \, ng \, ml^{-1}$ can be detected) is now achievable. Methods to measure cisplatin-induced DNA lesions include alkaline filter elution,[18] which detects DNA interstrand crosslinks, DNA–protein crosslinks (and DNA strand breaks); and ELISA techniques utilising polyclonal[19] or monoclonal[20] antibodies to detect various DNA intrastrand crosslinks. In addition, techniques have recently been described that measure binding and removal of DNA crosslinks in specific regions of the genome (*i.e.* transcribing *versus* non-transcribing genes).[21]

Identification of individual metabolites (and intact drug) may be determined using high-performance liquid chromatography (HPLC) separation and subsequent analysis by mass spectrometry. Studies have determined either 'total' platinum levels, which include drug bound to plasma proteins, or 'free' (ultrafilterable) levels. However, as platinum bound to plasma proteins is not cytotoxic,[22] pharmacokinetic studies have focused on the fate of free (ultrafilterable) platinum. Ultrafiltrates are usually prepared by centrifugation (1000 g, 30 min, 4 °C) using Amicon ultrafiltration cones (M_r cut-off of approximately 50 000).

The pharmacokinetics of cisplatin and carboplatin are clearly different and reflect largely the differences in chemical stability of the two drugs (Table 1). *In vitro* studies with plasma showed that, while cisplatin decomposes rapidly (half-life at 37 °C for binding to proteins of around 2 hours), carboplatin is more stable (half-life of 30 hours).[22] Cisplatin is inactivated primarily by avid binding to plasma proteins (> 90% is protein-bound 4 hours after infusion). Compared with cisplatin, carboplatin binds much less avidly to plasma proteins (around 24% protein-bound 4 hours after infusion) and is excreted primarily *via* the kidneys by glomerular filtration. Consequently, the majority of the plasma platinum is accounted for as the intact drug over the initial 4–6 hours after administration.[22]

Two independent studies have demonstrated an excellent relationship between renal function (as assessed by glomerular filtration rate, GFR) and the area under the plasma platinum concentration *versus* the time curve (AUC) for carboplatin (and hence the therapeutic efficacy and the severity of thrombocytopaenia).[23,24] These observations then led to Calvert and colleagues developing a simple dosing equation for carboplatin based on pretreatment kidney function:[23]

$$\text{Dose (mg)} = \text{AUC} \times (\text{GFR} + 25)$$

Use of this formula allows the adjustment of the dose of carboplatin according to renal function in order to produce optimal AUC values of $5\,\text{mg}\,\text{ml}^{-1}\,\text{min}^{-1}$ for previously treated patients and $7\,\text{mg}\,\text{ml}^{-1}\,\text{min}^{-1}$ for previously untreated patients.[23]

3 Platinum Chemistry

Platinum exists in two main oxidation states, Pt^{2+} and Pt^{4+}, usually designated Pt(II) and Pt(IV), respectively. In Pt(II) complexes such as cisplatin and carboplatin, the platinum atom has four bonds directed to the corners of a square plane at which the four ligand atoms are located. In contrast, in Pt(IV) complexes (such as iproplatin, tetraplatin, and JM216; see Figure 1), there are six bonds and ligands: four in a square-planar configuration, and two located axially, directly above and below the platinum, thus producing an octahedral configuration. The stereospecificity of the bonds is also of importance, as exemplified by the contrasting biological properties of the two isomers, cisplatin and transplatin (which is inactive).[25]

The chemical properties of platinum coordination complexes are largely dependent on the relative displacement reactions of the various ligands. While some platinum bonds (*e.g.* those to nitrogen or sulfur) are essentially irreversible under physiological conditions, the stability of bonds to halogens and especially to aquo (H_2O) is much lower. Thus, cisplatin reacts primarily by stepwise exchange of its two labile chlorides (leaving ligands) for water or hydroxyl ions (Figure 2). The final positively charged, highly reactive diaquo species (which is the same for cisplatin and carboplatin) is then capable of reacting with nucleophilic sites on DNA, RNA, or proteins. The presence of high chloride ion concentrations in extracellular fluid (approximately 100 mM) is considered to suppress the aquation reactions and allow the uncharged complex to penetrate

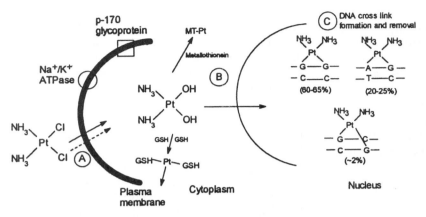

Figure 2 *Mechanisms of action and tumour resistance to cisplatin*

cell membranes. However, upon entering the cell, where the cytoplasmic chloride concentration is much lower (as low as 4 mM), the chloride ligands begin to exchange. Replacement of the two chloride ligands of cisplatin by the bidentate cyclobutane dicarboxylic acid (CBDCA) ligand in carboplatin results in a complex more than 100-fold as resistant to the above aquation reactions; in chloride-free phosphate buffer, pH 7, cisplatin had a half-life of 2.4 hours, compared with a half-life of 268 hours for carboplatin.[26]

Mechanism of Action

The cell-killing effects of cisplatin (and carboplatin) appear to be due to the formation of various stable bifunctional adducts on DNA, which then block replication or inhibit transcription. Supportive evidence for DNA as the critical target for the antitumour activity of cisplatin is provided by observations that cells from patients with diseases where DNA repair processes are deficient (*e.g.* Xeroderma pigmentosum) are hypersensitive to cisplatin.[27] In addition, correlations have been shown between levels of platinum–DNA adducts in peripheral blood cells (leukocytes) and disease response in patients receiving cisplatin or carboplatin.[28]

The nature and proportions of some of the various cisplatin-induced adducts on DNA are shown in Figure 2. These have been determined either by *in vitro* incubation of DNA and cisplatin or by extraction of DNA from cells exposed to cisplatin followed by separation and analysis of digested platinum-DNA adducts by HPLC.[29,30] The most common adduct (60–65%) involves binding of platinum to the nitrogen in position 7 of the imidazole ring of adjacent deoxyguanosines along the same strand of DNA (the GpG 1,2-intrastrand adduct). In addition, ApG (20–25%), GpXpG and ApXpG (5–6%; where X is any base) 1,3-intrastrand adducts, monofunctional adducts (2–3%), DNA–protein crosslinks, and G–G interstrand crosslinks (ISCs, 2%) are also found. Once bound to DNA, the adducts induced by cisplatin and carboplatin are similar in nature. The large difference in the concentrations required to produce equal numbers of adducts and similar cell-killing effects reflects the very much faster aquation rate for cisplatin.[26]

Perhaps surprisingly, the relative role of each of the various DNA adducts induced by cisplatin (especially intra- *versus* inter-strand adducts) in mediating cell-killing remains unclear. On one hand, there is supportive evidence emphasising the importance of 1,2-intrastrand adducts. The inactive transplatin is sterically unable to form the major GpG and ApG 1,2-intrastrand adducts formed by cisplatin. Instead a high proportion of DNA monoadducts are formed (up to 85% of adducts following a 1–2 hour drug incubation with DNA).[31] These are mainly detoxified through rapid reaction with glutathione. A minority slowly rearrange to form bifunctional 1,3 or 1,4 G–G intrastrand crosslinks or DNA interstrand crosslinks. Moreover, experiments have shown that the GpG intrastrand adduct is poorly repaired compared with GpXpG and the monofunctional adducts.[32,33]

Although they represent only approximately 2% of the total adducts formed by cisplatin, other findings suggest that ISCs may be important determinants of cytotoxicity. Early studies in L1210 leukaemia cells, again using *cis*- and *trans*-platinum, showed a relationship between cell killing and levels of ISCs.[34] Although both isomers produce ISCs in naked DNA, only cisplatin produces substantial ISCs in whole cells. Furthermore, studies determining repair at the level of individual genes in paired cisplatin-sensitive and -resistant human ovarian carcinoma cell lines have shown a marked increase in the gene-specific removal of ISCs in resistant lines, but no difference between sensitive and resistant lines in the removal of intrastrand adducts.[35,36] Recent data have also indicated that the nature of the ISCs formed by transplatin and by cisplatin in purified DNA differs, cisplatin favouring crosslink formation between guanines, and transplatin between guanine and complementary cytosines.[37]

Studies using both bacterial enzymes (*Escherichia coli* UVrABC nuclease[38] and mammalian cell extracts[39]) indicate that platinum–DNA adducts are removed from DNA by nucleotide excision repair. The *E. coli* UVrABC nuclease has been shown to incise the eighth phosphodiester bond 5′ and the fourth bond 3′ to G–G intrastrand crosslink, thereby excising an oligomer containing the adduct.[38] There is also considerable understanding of the molecular aspects of platinum–DNA interactions.[40] Interestingly, both the G–G and A–G intrastrand adducts unwind DNA by 13°, whereas the GXG 1,3-intrastrand adduct unwinds DNA by 23°; bending of the DNA double helix is similar (32–35°) for all three adducts.[40]

Structure-specific Damage-Recognition Proteins

Two classes of proteins (~ 28 kilodaltons and 80–100 kilodaltons) have been identified in mammalian cell extracts that bind specifically to the two major types of cisplatin-induced DNA 1,2-intrastrand crosslinks.[41,42] The proteins do not bind to adducts produced by the inactive transplatin.[42] A gene that encodes one protein of molecular weight of approximately 81 kilodaltons (termed structure-specific recognition protein, SSRP1) has been cloned and sequenced; it shares a region of homology with the high mobility group proteins, HMG1 and HMG2, the HMG box.[43] It has also been shown by two independent groups that HMG1 and HMG2 (proteins of 28.5 and 26.5 kilodaltons, respectively) can themselves recognise and bind to DNA platinated with cisplatin.[44,45]

The exact function of these damage recognition proteins, however, is currently unknown. Models have been proposed for their involvement in repairing cisplatin-damaged DNA or, somewhat to the contrary, in blocking access of repair enzymes to damaged DNA. Some support for the repair supposition is proved by observations that certain cisplatin-resistant cell tumour lines (some of which have an increase in DNA repair capacity) also exhibit an increase in damage recognition proteins.[46–48] In addition, a damage recognition protein complex, B1, has been shown to contain human single-stranded binding protein (HSSB),[49] a protein involved in an early stage of mammalian excision repair.[50] However, studies using other cell lines with resistance to cisplatin have found no obvious differences in levels of damage recognition proteins or in HSSB protein levels.[42,49,51,52] Furthermore, support for the model involving shielding of intrastrand crosslinks from DNA repair enzymes has arisen from studies in yeast, where a gene, lxr1, encoding a structure-specific recognition protein, when inactivated, resulted in a twofold decrease in sensitivity to cisplatin.[53]

4 Mechanisms of Resistance to Cisplatin/Carboplatin

Over recent years an intensive area of study has sought to determine mechanisms of tumour resistance (both intrinsic and acquired) to cisplatin/carboplatin. It is well established that resistance limits the clinical efficacy of the currently available platinum drugs, and thus an understanding of resistance mechanisms should aid the discovery of either improved platinum drugs or new strategies of modulating the efficacy of cisplatin/carboplatin. Studies have focused largely on the use of *in vitro* murine and human tumour cell lines and have revealed that resistance is often multifactorial in origin. It may arise through one or more of three major mechanisms: namely, decreased intracellular transport of drug, cytoplasmic detoxification through increased levels of thiol-rich species such as glutathione and/or metallothioneins, and enhanced removal of platinum-induced adducts from DNA and/or increased tolerance to platinum-DNA adducts (Figure 2: A, B, C).[54]

Many acquired cisplatin-resistant cell lines exhibit a decrease in platinum accumulation (typically 2- to 4-fold) as compared with their respective parent lines. However, the mechanisms by which platinum drugs enter cells are still unclear, with laboratory evidence supporting roles for both passive diffusion (*e.g.* the uptake of cisplatin in not saturable) and active transport (*e.g.* uptake can be modulated by a variety of pharmacological agents, including the Na^+K^+-ATPase inhibitor, ouabain). Notably, in contrast to the increased efflux of drug observed in tumour cells with multiple resistance to other commonly used anticancer drugs (*e.g.* doxorubicin, paclitaxel, etoposide and the Vinca alkaloids) mediated through 170 and/or 190 kilodalton membrane proteins, cisplatin resistance mediated at the level of the plasma membrane occurs mainly through reduced drug influx.[55]

As with reduced drug accumulation, many *in vitro* studies have produced evidence of a role for the cytoplasmic thiol-containing tripeptide glutathione (GSH) in mediating platinum drug resistance. In our own studies using eight human ovarian carcinoma cell lines exhibiting a 100-fold range in intrinsic sensitivity to cisplatin, GSH levels showed a significant correlation with cisplatin sensitivity; the most resistant line possesses GSH levels 4-fold higher than the most sensitive.[56] Evidence has also been reported for a direct interaction between cisplatin and GSH inside tumour cells, involving 2 mol of GSH complexed with 1 mol of platinum to form bis(glutathionato)platinum.[57] Elimination of this complex from tumour cells has been proposed to occur *via* an ATP-dependent glutathione S-conjugate export pump.[57] Cytoplasmic detoxification of platinum-based drugs may also occur through binding to metallothioneins, a class of cysteine-rich, low molecular weight isoproteins. Increased levels of metallothioneins have been reported in at least some acquired-cisplatin-resistant cell lines.[58]

The acquired resistance to cisplatin in many cell lines, including those in our own studies, appears to be due to enhanced repair/removal of platinum adducts from DNA.[54] Moreover, the possible relevance of repair in determining responses to platinum-based therapy in ovarian cancer patients has been highlighted in studies measuring RNA levels encoding for the ERCC1 human DNA repair gene.[59] Patients who were clinically resistant to cisplatin- (or carboplatin-) based therapy had expression levels of ERCC1 in their tumour biopsies 2.6-fold higher than those for patients who responded to that therapy. In addition, studies using tumour cells derived from testicular cancers indicate that they are hypersensitive to cisplatin.[60,61] There is some evidence to suggest that this hypersensitivity, which correlates with the clinical sensitivity of such tumours, might also be related to defective removal of platinum-DNA adducts.[62,63]

Other genes/proteins that may be involved in determining sensitivity to platinum drugs include various oncoproteins (*e.g.*, p21*RAS*, *MYC*, and *BCL2/BAX*), the p53 tumour suppressor gene, and a 60 kilodalton heat-shock protein (HSP60).[63] Transformation of mouse fibroblasts with *ras* oncogenes produced a 4- to 8-fold increase in cisplatin resistance.[64] Cisplatin has been shown to induce cell death through a programmed death pathway (apoptosis); at least some pathways of cell death appear to be activated through the wild-type p53 gene.[65] Interestingly, testicular tumours have been reported to exhibit p53 mutations only rarely, whereas tumours less responsive to chemotherapy commonly acquire p53 mutations.[66] Elevated constitutive levels of HSP60 have been observed in a human ovarian carcinoma cell line with acquired cisplatin resistance, as compared with its parent line.[67] Finally, modulation of various intracellular signal transduction pathways (*e.g.* those mediated by protein kinase C or *via* the epidermal growth factor) may also influence sensitivity and resistance to cisplatin.[68]

5   Circumvention of Tumour Resistance to Cisplatin

Although the discovery of carboplatin has unquestionably resulted in reduced morbidity for patients receiving platinum-based chemotherapy, there remains an urgent need for further improvements in clinical efficacy, particularly with regard to circumvention of tumour resistance. Four broad strategies may be envisaged for attempting to circumvent resistance: development of more broad-spectrum platinum drugs, administration of higher doses of cisplatin/carboplatin, combination chemotherapy with other active anticancer drugs, and combination with other pharmacological agents capable of modulating one or more of the known mechanisms of resistance to cisplatin.

Development of New Platinum Drugs

Despite the synthesis of many hundreds of cisplatin analogues over the past 20 years, there have been relatively few leads to the discovery of novel platinum drugs capable of circumventing tumour resistance to cisplatin/carboplatin.

Diaminocyclohexane (DACH) Platinum Complexes

Interest in platinum complexes based on the 1,2-diaminocyclohexane (DACH) carrier ligand [*e.g.* tetraplatin, tetrachloro-1,2-diaminocyclohexaneplatinum(IV), Ormaplatin; see Figure 1] arose from observations by Burchenal and colleagues, who showed that platinum complexes of this class retained activity in a cisplatin-resistant murine leukaemia (L1210) model.[69] Following chemical refinement (primarily to improve aqueous solubility), two DACH-based complexes, tetraplatin (ormaplatin) and oxaliplatin [oxalato-1,2-diaminocyclohexaneplatinum(II); 1-OHP] entered clinical trials.[70,71] Unfortunately, severe neurotoxicity appears to represent a major limitation with ormaplatin and to a lesser extent with oxaliplatin which has recently exhibited some activity in colorectal cancer when used in combination with 5-fluorouracil. In addition, an important *caveat* for adopting a single acquired-resistance murine tumour model, such as the L1210, for platinum drug development has been highlighted by studies where ormaplatin conferred no activity against another cisplatin-resistant murine model (the ADJ/PC6)[72] and exhibited poor activity against a variety of human ovarian carcinoma xenografts.[73]

Orally Active Platinum Complexes: JM216

The success of carboplatin calls attention to the importance of pursuing possibilities for improved patient quality of life in the treatment of cancer. As both cisplatin and carboplatin require intravenous administration, patient comfort could be further facilitated by use of an orally available platinum complex. A collaborative programme among the Johnson Matthey Technology Centre, Bristol Myers Squibb, and the Section of Drug Development at the Institute of Cancer Research (Sutton) has resulted in the discovery of a novel class of platinum(IV) complexes, the ammine/amine (or 'mixed amine') platinum(IV) dicarboxylates, which demonstrate promising oral activity against a variety of murine and human tumour preclinical models, as well as promising *in vitro* cytotoxic effects against human ovarian carcinoma cells exhibiting intrinsic resistance to cisplatin.[74,75] One such compound, [bis(acetato)(ammine)(dichloro)-cyclohexylamineplatinum(IV), JM216; Figure 1], entered clinical trials at the Royal Marsden Hospital (Sutton) in 1992 as the first orally administerable platinum drug.

Toxicology studies in rodents showed that JM216 possesses carboplatin-like, rather than cisplatin-like, properties, with myelosuppression dose-limiting and with no nephrotoxicity, hepatotoxicity, or neurotoxicity.[76–78] JM216 given orally showed comparable or superior activity to that observed for intravenous cisplatin or carboplatin against a variety of murine and human tumour models.[75] Antitumour effects appeared to be schedule-dependent, with greater activity being observed against a human ovarian carcinoma xenograft when a split daily dose (for five days every three weeks) was given, *versus* the same total dose given as a single bolus every 21 days.[78] Moreover, JM216 has exhibited preclinical circumvention of acquired cisplatin resistance, and especially of resistance mediated through reduced drug accumulation.[79,80]

JM216 has now undergone phase I clinical evaluation according to both single dose and daily × 5 schedules.[81,82] In agreement with the rodent data, myelosuppression has been the dose limiting toxicity, with no observations of nephro- or neurotoxicity. Worldwide phase II/III clinical trials are now ongoing.

Other Platinum-based Agents

Other platinum-based drugs currently undergoing early clinical evaluation are predominantly carboplatin-like, and have not, to date, shown convincing evidence of activity against cisplatin-resistant tumours.[54] Examples include CI-973 [*cis*-1,1-cyclobutanedicarboxylato(2R)-2-methyl-1,4-butanediamineplatinum(II); NK121], loboplatin [1,2-diaminomethylcyclobutaneplatinum(II)-lactate; D-19466], and DWA2114R [2-aminomethylpyrrolidine-(1,1-cyclobutanedicarboxylato)platinum(II) monohydrate].

Dose Intensification of Cisplatin/Carboplatin

In recent years, attempts have been made to increase the doses of cisplatin administered to patients, in the hope of achieving improved response rates. Although the renal and gastrointestinal toxicities of cisplatin may be ameliorated through intravenous hydration and forced diuresis, and 5-HT3 inhibitor antiemetics, respectively, disabling neurotoxicity has proven to be dose-limiting.[83] As the main toxicities of carboplatin are haematological, and dosage calculation based on kidney function are applicable,[23] carboplatin may be better suited to such studies (especially in combination with haematological support with growth factors).[84] However, it remains controversial whether increased doses will be reflected in improved patient survival. Our own data, using a panel of human ovarian carcinoma cell lines, have shown a difference in intrinsic cellular sensitivity to carboplatin and cisplatin of 30- and 100-fold, respectively.[85] Thus dose escalations of a similar magnitude may be necessary in the clinic.

Combinations with Other Anticancer Agents: Paclitaxel

The vast majority of currently available anticancer drugs do not exhibit significant activity against ovarian tumours that are, or have become, resistant to cisplatin, but in recent years, paclitaxel (taxol), a natural tubulin-binding compound extracted from the bark of the Pacific Yew, has shown promising levels of activity.[86] Response rates in the region of 20–30% have been reported. Clinical studies combining cisplatin and paclitaxel have been completed including a pivotal study in advanced ovarian cancer where the paclitaxel/cisplatin combination showed a marked improvement in survival compared to the conventional cyclophosphamide/cisplatin arm.[87]

Modulation of Platinum Resistance Mechanisms

To date, attempts at the modulation of platinum-resistance mechanisms have generally been conducted in preclinical models and have not reached the stage of clinical trials. A diverse range of agents, however, have shown promise,[88] and at least some may reach clinical trial in the near future. Examples include modulation at the level of the plasma membrane using the antifungal agent, amphotericin B; reduction of GSH levels using buthionine sulfoximine; and inhibition of DNA repair using aphidicolin.

6   Summary

Over the past 25 years, the platinum-based drugs (cisplatin and, latterly, the less toxic analogue carboplatin) have brought significant therapeutic benefit to a large number of cancer sufferers. However, there remains scope for substantial improvement in the clinical utility of platinum-based metal coordination complexes, either through the discovery of additional platinum-based complexes possessing a wider therapeutic spectrum of activity (see Chapter 8) or through pharmacological modulation of platinum-resistance mechanisms. The likelihood of achieving these important goals has undoubtedly been enhanced through our vastly increased understanding of the chemistry and the biochemical properties of cisplatin.

References

1 B. Rosenberg, *Cancer (Phila).*, 1985, **55**, 2303.
2 L.H. Einhorn and J.D. Donohue, *Annals Intern. Med.*, 1977, **87**, 293.
3 E. Wiltshaw and B. Carr, in *Platinum, Coordination Complexes in Cancer Chemotherapy*, ed. T.A. Connors and J.J. Roberts, Springer-Verlag, Heidelberg, 1974, p. 178.

4 J.P. Neijt, W.W. ten Bokkel Huinink, M.E.L. van der Burg, A.T. van Oostersom, P.H.B. Willemse, J.B. Vermorken, A.C.M van Lindert, A.P.M. Heintz, E. Aartsen, M. van Lent, J.B. Trimbos and A.J. de Meijer, *Eur. J. Cancer*, 1991, **11**, 1367.

5 P.J. Loehrer and L.H. Einhorn, *Annals Intern. Med.* 1984, **100**, 704.

6 Advanced Ovarian Cancer Trialists Group, *Br. Med. J.*, 1991, **303**, 884.

7 C.J. Williams, L.Stewart, M. Parmar and D. Guthrie, *Semin. Oncol.*, 1992, **19** (Suppl 2), 120.

8 A.E. Taylor, E. Wiltshaw, M. Gore, I. Fryatt and C. Fisher, *J. Clin. Oncol.*, 1994, **12**, 2066.

9 E. Eisenhauer, K. Swerton, J. Sturgeon, S. Fine, S. O'Reilly and R. Canetta, in *Carboplatin; Current Perspectives and Future Directions*, ed. P. Bunn, R. Canetta, R. Ozols and M. Rozencweig, W.B. Saunders Company, Philadelphia, 1990, p. 133.

10 M. Gore, I. Fryatt, E. Wiltshaw, T. Dawson, B. Robinson and A. Calvert, *Br. J. Cancer*, 1989, **60**, 767.

11 M. Markman, R. Rothman, T. Hakes, B. Reichman, W. Hoskins, S. Rubin, W. Jones, L. Almadrones and J.L. Lewis Jr., *J. Clin. Oncol.*, 1991, **9**, 289.

12 D.S. Alberts, R. Canetta and N. Mason-Liddil., *Semin. Oncol.* 1990, **17**, 54.

13 B.J. Foster, B.J. Harding, M.K. Wolpert-DeFilippes, L.Y. Rubinstein, K. Clagett-Carr and B. Leyland-Jones, *Cancer Chemother. Pharmacol.*, 1990, **25**, 395.

14 C. Sessa, J. Vermorken, J. Renard, S. Kaye, D. Smith, W. ten Bokkel Huinink, F. Cavalli and H. Pinedo, *J. Clin. Oncol.*, 1988, **6**, 98.

15 G. Weiss, S. Green, D.S. Alberts, J.T. Thigpen, H.E. Hines, K. Hanson, H.I. Pierce, L.H. Baker and J.W. Goodwin., *Eur. J. Cancer*, 1991, **27**, 135.

16 C. Trask, A. Silverstone, C.M. Ash, H. Earl, C. Irwin, A. Bakker, J.S. Tobias and R.L. Souhami, *J. Clin. Oncol.* 1991, **9**, 1131.

17 W.P. McGuire III, J. Arseneau, J.A. Blessing, P.J.DiSaia, K.D. Hatch, F.T. Given, Jr., N.N.H. Teng and W.T. Creasman, *J. Clin. Oncol.*, 1989, **7**, 1462.

18 K.W. Kohn, R.A.G. Ewig, L.C. Erickson and L.A. Zwelling, in *Handbook of DNA repair*, ed. E.C. Friedberg and P.C. Hanawalt, Marcel Dekker Inc., New York, 1981, p. 379.

19 M.C. Poirier, S.J. Lippard, L.A. Zwelling, H.M. Ushay, D. Kerrigan, C.C. Thill, R.M. Santella, D. Grunberger and S.H. Yuspa, *Proc. Natl. Acad. Sci. USA*, 1982, **79**, 6443.

20 M.J. Tilby, C. Johnson, R.J. Knox, J. Cordell, J.J. Roberts and C.J. Dean, *Cancer Res.*, 1991, **51**, 123.

21 J.C. Jones, W. Zhen, E. Reed, R.J. Parker, A. Sancar and V.A. Bohr, *J. Biol. Chem.* 1991, **266**, 7101.

22 S.J. Harland, D.R. Newell, Z. H. Siddik, R. Chadwick, A.H. Calvert and K.R. Harrap, *Cancer Res.*, 1984, **44**, 1693.

23 A.H. Calvert, D.R. Newell, L.A. Gumbrell, S. O'Reilly, M. Burnell, F.E. Boxall, Z.H. Siddik, I.R. Judson, M.E. Gore and E. Wiltshaw, *J. Clin. Oncol.*, 1989, **7**, 1748.

24 M.J. Egorin, D.A. Van Echo, S.J. Tipping, E.A. Olman, M.Y. Whitacre, B.W. Thompson and J. Aisner, *Cancer Res.*, 1984, **44**, 5432.

25 T.A. Connors, M.J. Cleare and K.R. Harrap, *Cancer Treat. Rep.*, 1979, **63**, 1499.

26 R.J. Knox, F. Friedlos, D.A. Lydall and J.J. Roberts, *Cancer Res.*, 1986, **46**, 1972.

27 J.J. Roberts, R.J. Knox, F. Friedlos and D.A. Lydall, in *Biochemical Mechanisms of Platinum Antitumour Drugs*, ed. D.C.H. McBrien and T.F. Slater, IRL Press, Oxford, 1986, p. 29.

28 E. Reed, R.F. Ozols, R. Tarone, S.H. Yuspa and M.C. Poirier, *Proc. Natl. Acad. Sci. USA*, 1987, **84**, 5024.

29 A. Eastman, *Biochemistry*, 1986, **25**, 3912.

30 A.M.J. Fichtinger-Schepman, J.L. van der Veer, J.H.J. den Hartog, P.H.M. Lohman and J. Reedijk, *Biochemistry*, 1985, **24**, 707.
31 A. Eastman and M.A. Barry, *Biochemistry*, 1987, **26**, 3303.
32 J.D. Page, I. Husain, A. Sancar and S.G. Chaney, *Biochemistry*, 1990, **29**, 1016.
33 D.E. Szymkowski, K. Yarema, J.M. Essigmann, S.J. Lippard and R.D. Wood, *Proc. Natl. Acad. Sci. USA.*, 1992, **89**, 10 772.
34 L.A. Zwelling, T. Anderson and K.W. Kohn, *Cancer Res.*, 1979, **39**, 365.
35 W. Zhen, C.J. Link, P.M. O'Connor, E. Reed, R. Parker, S.B. Howell and V.A. Bohr, *Mol. Cell Biol.*, 1992, **12**, 3689.
36 S.W. Johnson, R.P. Perez, A.K. Godwin, A.T. Yeung, L.M. Handel, R.F. Ozols and T.C. Hamilton, *Biochem. Pharmacol.*, 1994, **47**, 689.
37 V. Brabec and M. Leng, *Proc. Natl. Acad. Sci. USA.*, 1993, **90**, 5345.
38 D.J. Beck, S. Popoff, A. Sancar and W.D. Rupp, *Nucl. Acid Res.*, 1985, **13**, 7395.
39 J. Hansson and R.D. Wood, *Nucl. Acid Res.*, 1989, **17**, 8073.
40 K.M. Comess and S.J. Lippard, in *Anticancer Drug-DNA Interactions*, ed. S. Neidle and M. Waring, MacMillan Press Ltd, London, 1993, vol. 1, p. 134.
41 G. Chu and E. Chang, *Science*, 1988, **242**, 564.
42 B.A. Donahue, M. Augot, S.F. Bellon, D.K. Treiber, J.H. Toney, S.J. Lippard and J.M. Essigmann, *Biochemistry*, 1990, **29**, 5872.
43 S.L. Bruhn, P.M. Pil, J.M. Essigmann, D.E. Housman and S.J. Lippard, *Proc. Natl. Acad. Sci. USA.*, 1992, **89**, 2307.
44 E.N. Hughes, B.N. Engelsberg and P.C. Billings, *J. Biol. Chem.*, 1992, **267**, 13 520.
45 P.M. Pil and S.J. Lippard, *Science*, 1992, **256**, 234.
46 G. Chu and E. Chang, *Proc. Natl. Acad. Sci. USA.*, 1990, **87**, 3324.
47 C.C.-K. Chao, S.-L. Huang, L.-Y. Lee and S. Lin-Chao, *Biochem. J.*, 1991, **277**, 875.
48 K. McLaughlin, G. Coren, J. Masters and R. Brown, *Int. J. Cancer*, 1993, **53**, 662.
49 C.K. Clugston, K. McLaughlin, M.K. Kenny and R. Brown, *Cancer Res.*, 1992, **52**, 6375.
50 D. Coverley, M.K. Kenny, M. Munn, W.D. Rupp, D.P. Lane and R.D. Wood, *Nature*, 1991, **349**, 538.
51 P.A. Andrews and J.A. Jones, *Cancer Commun.*, 1991, **3**, 1.
52 D. Bissett, K. McLaughlin, L.R. Kelland and R. Brown, *Br. J. Cancer*, 1993, **67**, 742.
53 S.J. Brown, P.J. Kellett and S.J. Lippard, *Science*, 1993, **261**, 603.
54 L.R. Kelland, *Crit. Rev. Oncol/Hematol.*, 1993, **15**, 191.
55 D.P. Gately and S.B. Howell, *Br. J. Cancer*, 1993, **67**, 1171.
56 P. Mistry, L.R. Kelland, G. Abel, S. Sidhar and K.R. Harrap, *Br. J. Cancer*, 1991, **64**, 215.
57 T. Ishikawa and F. Ali-Osman, *J. Biol. Chem.*, 1993, **268**, 20 116.
58 S.L. Kelley, A. Basu, B.A. Teicher, M.P. Hacker, D.H. Hamer and J.S. Lazo, *Science.*, 1988, **241**, 1813.
59 M. Dabholkar, F. Bostick-Bruton, C. Weber, V.A. Bohr, C. Egwuagu and E. Reed, *J. Natl. Cancer Inst.*, 1992, **84**, 1512.
60 M.F. Pera, F. Friedlos, J. Mills and J.J. Roberts, *Cancer Res.*, 1987, **47**, 6810.
61 L.R. Kelland, P. Mistry, G. Abel, F. Friedlos, S.Y. Loh, J.J. Roberts and K.R. Harrap, *Cancer Res.*, 1992, **52**, 1710.
62 B.T. Hill, K.J. Scanlon, J. Hansson, A. Harstrick, M. Pera, A.M.J. Fichtinger-Schepman and S.A. Shellard, *Eur. J. Cancer*, 1994, **30A**, 832.
63 L.R. Kelland, *Eur. J. Cancer*, 1994, **30A**, 725.
64 S. Isonishi, D.K. Hom, F.B. Thiebaut, S.C. Mann, P.A. Andrews, A. Basu, J.S. Lazo, A. Eastman and S.B. Howell, *Cancer Res.*, 1991, **51**, 5903.

65 S.W. Lowe, H.E. Ruley, T. Jacks and D.E. Housman, *Cell*, 1993, **74**, 957.

66 H.-Q. Peng, D. Hogg, D. Malkin, D. Bailey, B.L. Gallie, M. Bulbul, M. Jewett, J. Buchanan and P.E. Goss, *Cancer Res.*, 1993, **53**, 3574.

67 E. Kimura, R.E. Enns, F. Thiebaut and S.B. Howell, *Cancer Chemother. Pharmacol.*, 1993, **32**, 279.

68 S.B. Howell, S. Isonishi, R.C. Christen, P.A. Andrews, S.C. Mann and D. Hom, in *Platinum and Other Metal Coordination Compounds in Cancer Chemotherapy*, ed. S.B. Howell, Plenum Press, New York, 1991, p. 173.

69 J.H. Burchenal, K. Kalaher, K. Dew and L. Lokys, *Cancer Treat. Rep.*, 1979, **63**, 1493.

70 R.J. Schilder, F.P. LaCreta, R.P. Perez, S.W. Johnson, J.M. Brennan, A. Rogatko, S. Nash, C. McAleer, T.C. Hamilton, D. Roby, R.C. Young, R.F. Ozols and P.J. O'Dwyer, *Cancer Res.*, 1994, **54**, 709.

71 D. Machover, E. Diaz-Rubio, A. de Gramont, A. Schilf, J.-J. Gastiaburu, S. Brienza, M. Itzhaki, G. Metzger, D. N'Daw, J. Vignoud, A. Abad, E. Francois, E. Gamelin, M. Marty, J. Sastre, J.-F. Seitz and M. Ychou, *Annals Oncol.*, 1996, **7**, 95.

72 P.M. Goddard, M.R. Valenti and K.R. Harrap, *Annals Oncol.*, 1991, **2**, 535.

73 M. Jones, J. Siracky, L.R. Kelland and K.R. Harrap, *Br. J. Cancer.*, 1993, **67**, 24.

74 L.R. Kelland, B.A. Murrer, G. Abel, C.M. Giandomenico, P. Mistry and K.R. Harrap, *Cancer Res.*, 1992, **52**, 822.

75 L.R. Kelland, G. Abel, M. McKeage, M. Jones, P.M. Goddard, M. Valenti, B.A. Murrer and K.R. Harrap, *Cancer Res.*, 1993, **53**, 2581.

76 M.J. McKeage, S.E. Morgan, F.E. Boxall, B.A. Murrer, G.C. Hard and K.R. Harrap, *Br. J. Cancer*, 1993, **67**, 996.

77 M.J. McKeage, F.E. Boxall, M. Jones and K.R. Harrap, *Cancer Res.*, 1994, **54**, 629.

78 M.J.McKeage, L.R. Kelland, F.E. Boxall, M.R. Valenti, M. Jones, P.M. Goddard, J. Gwynne and K.R. Harrap, *Cancer Res.*, 1994, **54**, 4118.

79 K.J. Mellish, L.R. Kelland and K.R. Harrap, *Br. J. Cancer*, 1993, **68**, 240.

80 S.Y. Loh, P. Mistry, L.R. Kelland, G. Abel, B.A. Murrer and K.R. Harrap, *Br. J. Cancer*, 1992, **66**, 1109.

81 M.J. McKeage, P. Mistry, J. Ward, F.E. Boxall, S. Loh, C. O'Neill, P. Ellis, L.R. Kelland, S.E. Morgan, B. Murrer, P. Santabarbara, K.R. Harrap and I.R. Judson, *Cancer Chemother. Pharmacol.*, 1995, **36**, 451.

82 M.J. McKeage, F. Raynaud, J. Ward, C. Berry, D. Odell, L.R. Kelland, B. Murrer, P. Santabarbara, K.R. Harrap and I. R. Judson. *J. Clin. Oncol.*, 1997, **15**, 2691.

83 R.F. Ozols, Y. Ostchega, C.E. Myers and R.C. Young, *J. Clin. Oncol.*, 1985, **3**, 1246.

84 A.H. Calvert, D.R. Newell and M.E. Gore, *Semin. Oncol.*, 1992, **19**, 155.

85 C.A. Hills, L.R. Kelland, G. Abel, J. Siracky, A.P. Wilson and K.R. Harrap, *Br. J. Cancer*, 1989, **59**, 527.

86 E.L. Trimble, J.D. Adams, D. Vena, M.J. Hawkins, M.A. Friedman, J.S. Fisherman, M.C. Christian, R. Canetta, N. Onetto, R. Hayn and S.G. Arbuck, *J. Clin. Oncol.*, 1993, **11**, 2405.

87 W.P. McGuire, W.J. Hoskins, M.F. Brady, P.R. Kucera, E.E. Partridge, K.Y. Look, D.L. Clarke-Pearson and M. Davidson, *New Engl. J. Med.*, 1996, **334**, 1.

88 H. Timmer-Bosscha, N.H. Mulder and E.G.E. de Vries, *Br. J. Cancer*, 1992, **66**, 277.

CHAPTER 8

Dinuclear and Trinuclear Platinum Anticancer Agents

NICHOLAS FARRELL[1] AND SILVANO SPINELLI[2]

[1]Department of Chemistry, Virginia Commonwealth University, 1001 W. Main Street, Richmond, VA 23284, USA
[2]Boehringer Mannheim Italia, Research Division, Via Monza, Monza, Italy

1   Introduction

In early 1998 a novel trinuclear platinum compound, BBR3464, entered Phase 1 clinical trials, the first genuinely new platinum agent not based on the 'classical' cisplatin structure to do so. Its structure is based on the dinuclear concept developed in our laboratories and is best described as two $[PtCl(NH_3)_2]$ units linked by a non-covalent tetra-amine $[Pt(NH_3)_2\{H_2N(CH_2)_6NH_2\}_2]$ (Figure 1). A 1994 World Health Organization consultation document recognized 24 drugs as essential for rational management of advanced malignant neoplasms. Of these, cisplatin, $[cis\text{-}[PtCl_2(NH_3)_2]$, cis-DDP] is cited for treatment of germ-cell cancers, gestational trophoblastic tumors, epithelial ovarian cancer and small cell lung cancer as well as for palliation of bladder, cervical, nasopharyngeal, oesophageal and head and neck cancers. The use of cisplatin (usually as a principal component of combination regimens) has rendered at least one cancer, testicular cancer, curable and is significant in treatment of ovarian and bladder cancers. Despite considerable effort the consensus is that 'second-generation' structural analogues such as carboplatin ($[Pt(CBDCA)\text{-}(NH_3)_2)]$, CBDCA = 1,1-cyclobutanedicarboxylate) do not indicate, disappointingly, a broader spectrum of activity in comparison with the parent drug.[1]

All direct structural analogs of cisplatin produce a very similar array of adducts on target DNA and it is therefore not surprising that they induce similar biological consequences. This latter consideration led us to challenge the empirical structure–activity relationships and formulate the hypothesis that development of platinum compounds structurally *dissimilar* to cisplatin may, by virtue of formation of different types of Pt–DNA adducts, lead to compounds with a spectrum of clinical activity genuinely complementary to the parent drug.[2,3] We considered that future discovery of clinically useful platinum agents was likely to

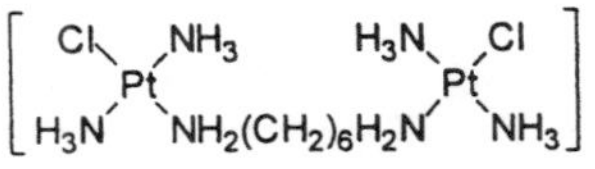

$$\left[\begin{array}{cc} Cl & NH_3 \\ \diagdown Pt \diagup & \\ H_3N & NH_2(CH_2)_6H_2N \end{array} \quad \begin{array}{cc} H_3N & Cl \\ \diagdown Pt \diagup & \\ & NH_3 \end{array} \right] (NO_3)_2$$

1,1/t,t

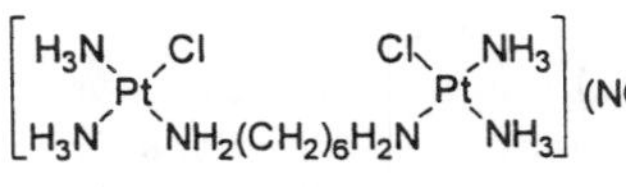

$$\left[\begin{array}{cc} H_3N & Cl \\ \diagdown Pt \diagup & \\ H_3N & NH_2(CH_2)_6H_2N \end{array} \quad \begin{array}{cc} Cl & NH_3 \\ \diagdown Pt \diagup & \\ & NH_3 \end{array} \right] (NO_3)_2$$

1,1/c,c

$$\left[\begin{array}{cc} Cl & NH_3 \\ \diagdown Pt \diagup & \\ H_3N & NH_2(CH_2)_6H_2N \end{array} \quad \begin{array}{cc} H_3N & NH_2(CH_2)_6H_2N \\ \diagdown Pt \diagup & \\ & NH_3 \end{array} \quad \begin{array}{cc} NH_3 \\ \diagdown Pt \diagup \\ H_3N & Cl \end{array} \right] (NO_3)_4$$

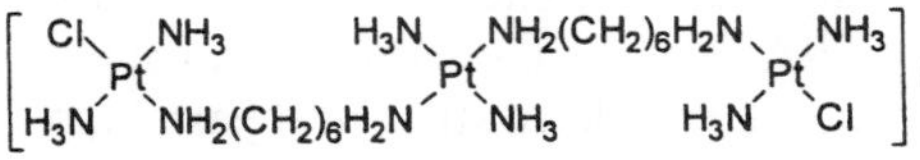

BBR 3464 (1,0,1/t,t,t)

Figure 1 *Structure for the trinuclear clinical drug BBR3464 and its dinuclear analogues. For a convenient abbreviation of polynuclear platinum complexes we have adopted a system where the numbers refer to the number of chlorides (or anionic leaving groups) on each platinum atom. Where there are two chlorides on the same Pt, the lettering specifies their mutual geometries (cis or trans). For those possibilities where there is only one chloride in a coordination sphere, the lettering refers to the geometry with respect to the nitrogen of the bridging diamine. Once these two parameters are specified, the geometry of the overall complex is automatically fixed. Thus [{trans-PtCl(NH₃)₂}₂H₂N(CH₂)ₙNH₂]Cl₂ is 1,1/t,t etc. Trinuclear agents are abbreviated in the same manner – thus BBR3464 contains one Cl on the end platinum units while the central unit is a tetra-amine which contains no (zero) Cl and is incapable of forming covalent bonds. Since the Cl atoms are trans to the diamine bridge the abbreviation is 1,0,1/t,t,t*

arise with 'non-classical' structures of which the subject of this chapter, dinuclear and trinuclear bifunctional DNA-binding agents, are amongst the best studied.

2 Dinuclear and Trinuclear Platinum Complexes as Anticancer Agents

During the last ten years, we have prepared a number of novel dinuclear Pt compounds, where two platinum-amine coordination units are linked by a variable length diamine linker, as potential antitumor drugs. We and our collaborators have extensively studied these new classes of platinum compounds in order to understand the patterns of DNA modification induced by various structural motifs and further relate these patterns to cytotoxicity and antitumour activity.[4,5] This has been successful to the point where one compound has now advanced to full clinical testing. The dinuclear structure is extremely flexible and capable of producing a wide series of compounds differing in functionality (bifunctional to tetrafunctional DNA-binding), geometry (leaving chloride groups *cis* or *trans* to the diamine bridge), non-leaving groups in the Pt coordination spheres (NH_3 or a planar group such as pyridine or quinoline) and linker (flexible, variable chain length). Our initial synthetic efforts identified the 1,1/t,t series (see Figure 1 for abbreviations) as having the most promising pattern of antitumor activity and DNA-binding. As synthetic efforts progressed, we envisaged complexes with improved DNA affinity for long-range cross-linking. This was accomplished first through the incorporation of a third platinum center within the alkane diamine chain (Figure 1). The structure, which arose from our initial reports on trinuclear systems,[6] is notable for the presence of the central Pt which contributes to DNA affinity only through electrostatic and H-bonding interactions. The $4+$ charge, the presence of at least two Pt coordination units capable of binding to DNA and the consequences of such DNA binding is a remarkable departure from the cisplatin structural paradigm and argue that the drug should be considered the first representative of an entirely new structural class of anticancer agents. With this advance the paradigm of cisplatin-based antitumor agents is altered.

3 Biological Activity of Polynuclear Platinum Complexes. Summary and p53 Status of Human Tumors Treatable by BBR3464

The profile of preclinical activity of BBR3464 mirrors its unique structure and is characterized by activity in human tumor (*e.g.* ovarian) xenografts resistant to cisplatin and alkylating agents; a high activity in a broad spectrum of human tumors commonly insensitive to chemotherapeutic intervention (*e.g.* non-small-cell lung, gastric) and characterized as p53-mutant. In this section we compare briefly the antitumor activity of the lead clinical candidate with cisplatin in a range of human tumor xenografts. The biological testing has been performed in collaboration with Boehringer Mannheim Italia, now Roche-Boehringer Mannheim. The general approach has been (i) to obtain cytotoxicity and *in vivo* data in murine L1210 and human ovarian A2780 models both sensitive and resistant to cisplatin; (ii) to extend the most promising compounds to testing in other solid tumor murine (M5076) and human solid tumor (IGROV-1 (ovarian), LX-1 (small cell lung) models and (iii) finally extended testing on a range of ovarian, lung, colon and gastric xenografts. This approach has the advantage of obtaining appropriate pharmacological data on effective and toxic doses on the murine L1210 animals and on the A2780 human ovarian animals, where there is considerable historical precedent for cisplatin in both cases. Table 1 summarizes the overall profile from 18 human tumors tested.[7] In almost all cases BBR3464 is more effective than cisplatin. Noteworthy also is the optimal dose of $0.3\,\mathrm{mg/kg^{-1}}$ for BBR3464 – allowing for molecular weight differences, the new agent is approximately 100-fold more effective than cisplatin for the maximum therapeutic effect.

Biological Activity and p53 Status

A highly relevant area of research interest is the p53 gene status of human tumors. The p53 protein is recognized as an important cell regulatory element necessary for cell cycle arrest and apoptosis induction. Wild type p53 function

Table 1 *Comparision at maximum tolerated dose of BBR3464* $(0.2–0.4\,mg\,kg^{-1})$ *and cisplatin* $(3–6\,mg\,kg^{-1})$ *after i.v. repeated treatment on staged tumors*

Clinical parameter[a]	BBR 3464	Cisplatin
Resistance, TWI < 50%	0	9 (4 NSCLC, 2 ovarian, 2 gastric, 1 prostatic)
Relative resistance, TWI 50–70%	3 (1 NSCLC, 1 gastric, 1 prostatic)	7 (2 SCLC, 2 NSCLC, 2 ovarian, 1 bladder)
Sensitivity, TWI > 70%	15 (3 SCLC, 5 NSCLC, 5 ovarian, 1 gastric,1 bladder)	2 (1 ovarian, 1 SCLC)

[a]See reference 7. TWI% is Tumor Weight Inhibition compared with controls. SCLC: small cell lung cancer; NSCLC: non-small cell lung cancer. The clinical parameter refers to the fact that clinical resistance, relative resistance and sensitivity are most likely to be seen at these TWI levels. Thus, BBR3464 is significantly more potent than cisplatin – good tumor sensitivity was observed in 15/18 cases for BBR3464.

plays a critical role in cellular response to drug-induced DNA damage, although there is no simple relationship between DNA damage and p53 functional status.[8] Loss of normal p53 function could lead to intrinsic drug resistance as a result of reduced cell ability to engage an apoptotic response. In addition to a role in modulating apoptosis, p53-mediated cellular responses include cell growth arrest required to allow DNA repair prior to genome replication. The protein is further implicated in playing a role in modulating nucleotide excision repair pathways. A subset of the tumors was evaluated by Dr. F. Zunino for p53 status.[9] The new agent displays remarkably high activity in mutant p53 tumors (Table 2). In 5 out of 6 cases the TWI is indicative of clinical sensitivity. In the sixth, a prostatic tumor, the reponse is still significantly better than *cis*-DDP. So, not only can this new clinical candidate be presented as active in tumors with both acquired and inherent resistance to *cis*-DDP but it can be presented as maintaining activity in tumors with mutant p53. The plausible explanation of the hypersensitivity of human tumors with mutant p53 to BBR3464 is that apoptosis induced by the drug is not mediated by p53. A plausible extension of this thinking, then, is that the novel DNA-binding modes of BBR3464 are not recognised by p53 and so repair and/or apoptotic pathways are not induced.

Comparison with Other Clinical Cross-linking Agents

BBR3464 clearly represents a significantly promising antitumor agent and a distinct departure from all previous mononuclear cisplatin agents. This is also true in comparison with other DNA–DNA cross-linking agents. Table 3 compares activity with the clinically used alkylating agents. In these three cases, chosen because of strict ability to compare, BBR3464 is equivalent in efficacy to cisplatin in one and significantly better in the other and significantly more potent than the alkylating agents with the one exception of Mitomycin C in LXFE839.

Table 2 *Activity and p53 status of human tumours treated with platinum compounds.*[a]

Tumor model/type	Drug[b] mg kg^{-1}	TWI%	Resistance	p53 Status
POVD/DDP(SCLC)	Cisplatin 6	70	Acquired	*Mutant*
	3464 0.4	93		
Calu-3(NSCLC)	Cisplatin 6	60	Intrinsic	*Mutant*
	3464 0.3	92		
LX-1(NSCLC)	Cisplatin 4	38	Intrinsic	*Mutant*
	3464 0.3	73		
POCS	Cisplatin 6	56	Intrinsic	*Mutant*
(SCLC)	3464 0.4	92		
IGROV/DDP	Cisplatin 6	68	Acquired	*Mutant*
(Ovarian)	3464 0.3	80		

[a]Adapted from reference 9. Tumor fragments s.c. into CD1 nu/nu female mice (day 0). Treatment i.v. on days 1, 8, 15 when tumor weight reached an average of 100 mg. All compounds were dissolved in saline before use. TWI%: tumor weight inhibition (on day 22): 100 − (mean relative tumor weight of treated mice/mean relative tumor weight of controls × 100). Relative tumour weight is determined as T_x/T_o where T_o is the tumor weight at the start of treatment (day 1) and T_x is the tumor weight at day x.

Table 3 *Comparison of efficacy in three solid tumors (TWI%) of platinum agents with clinically used DNA-DNA crosslinking agents*

Agent	Dose mg kg^{-1} day^{-1}	Schedule, Route	LXFE839	LXFA526	OVXF899
Cyclophosphamide	200	1, 15(i.p.)	31	43	21
Ifosfamide	130.0	1, 15(i.p.)	—	46	34
Mitomycin	2.0	1, 15(i.v.)	85	23	53
BBR3464	0.3	1, 8, 15(i.v.)	62	51	75
Cisplatin	3.0	1, 8, 15(i.v.)	55	23	30

Data and comparison supplied by Dr. H.H. Fiebig, Klinikum der Albert-Ludwigs-Universität Freiburg through Boehringer Mannheim Italia. All tumors transplanted s.c. TWI is (1 − tumor area T/C) × 100. LXFE: epidermoid lung cancer; LXFA: lung cancer; OVXF: ovarian cancer.

4   Structure–Activity Relationships in Polynuclear Platinum Complexes

An immediate question relating to the structure–activity relationships within this class of compounds is immediately raised – whether it is necessary to contain a third Pt unit (trinuclear class of compounds) in the molecule to achieve this remarkable antitumor potency. The most closely related compounds to BBR3464 would contain a diamine backbone with some hydrogen-bonding capacity. This was achieved through synthesis of dinuclear platinum complexes with hydrogen-bonding ligands spermine (total charge 4 +) and spermidine (total charge 3 +)

linkers (Figure 2).[10] Selective platination to the terminal NH_2 group is achieved by selective blocking and deblocking of the secondary nitrogens.[11] Upon complexation, the Boc group may be displaced readily by dilute acid.

Designed synthesis of polyamine backbones in this manner mimics the essential features of BBR3464 such as charge and chain length – although chain length is not absolutely critical but the extra charge is. The shorter spermidine compound is also very potent. Interestingly, the absence of the charged central NH_2 groups reduces cytotoxicity considerably (compare the spermidine pair BBR3571 and BBR3537 where the Boc is still present on the central nitrogens) and the Boc loses its activity in a resistant cell line. Their biological activity in human tumors such as LX-1 reproduces that of BBR3464 and is nothing short of exceptional and significantly improved over the 'original' straight-chain diamines such as BBR3005 (Table 4). The results show that a family of dinuclear compounds similar to BBR3464 can be achieved through use of polyamine linkers. Just as a molecule of structure [*cis*-PtX$_2$(amine)$_2$] is now expected to be antitumor active with predictable DNA binding characteristics [high proportion of d(GpG) intrastrand cross-links, few interstrand cross-links, bending of DNA into major groove], di- and tri-nuclear complexes incorporating the principal features of *trans*-Pt monofunctional coordination spheres, separated by a long-chain, flexible linker capable of hydrogen-bonding and with an overall charge of 3 + or 4 + are now expected to reproduce high antitumor activity, activity in p53 mutant tumours and a predictable DNA binding profile.

5 DNA Binding of Polynuclear Platinum Complexes

The interactions of bifunctional polynuclear platinum complexes with target DNA are unlike those of any DNA-damaging agent in clinical use. This program developed from the hypothesis that DNA damage different to cisplatin will lead to a profile of antitumor activity that is also different and, by definition, complementary in the clinic. The question remains – how are they connected? The mechanistic implications must be that the range of adducts produced by the new agents are sufficiently unique that DNA processing and repair pathways are distinctly different from those of any other agents. The mechanistic importance of the polynuclear platinum drugs is that we tend to consider classes of DNA-damaging agents as acting by different global mechanisms – cross-link formation (cisplatin/alkylating agents); strand breakage (bleomycin); topoisomerase inhibition (VP-16, anthracyclines). In the case of the polynuclear class of compounds we consider that cross-link formation is also the major lesion but these are sufficiently different in structure and consequence to previously studied covalent adducts that the cells' response follows different pathways. The ability to modify biological response by suitable coordination chemistry on DNA could be a significant contribution of inorganic chemistry to medicine.

The DNA binding profile of BBR3464 is similar to that described for the dinuclear agents (Table 5). We have summarized this profile of DNA binding in recent reviews.[2,3] The formally bifunctional DNA binding is distinctly different to that of cisplatin and is highlighted by a high percentage of interstrand

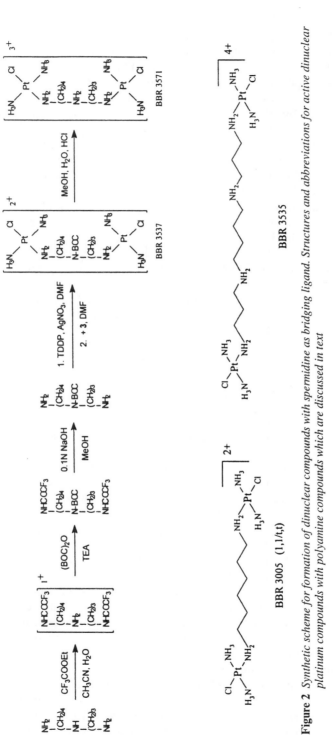

Figure 2 *Synthetic scheme for formation of dinuclear compounds with spermidine as bridging ligand. Structures and abbreviations for active dinuclear platinum compounds which are discussed in text*

Table 4 *Cytotoxicity and antitumor activity of platinum polyamine complexes in murine leukemia sensitive (L1210) and resistant to CDDP (L1210/CDDP) and in the LX-1 human tumor xenograft*[a]

Compound	*In vitro* IC_{50}, (μM) L1210	*In vitro* IC_{50}, (μM) L1210/ CDDP	*In vivo (L1210/CDDP only)*[c] Dose (mg kg^{-1} day^{-1})	*In vivo (L1210/CDDP only)*[c] T/C%	*In vivo LX-1* Dose (mg kg^{-1} day^{-1})	*In vivo LX-1* TWI%
BBR 3535	0.63	1.26	1	161, 178	1	88
BBR 3537	0.88	4.64	2	100	—	—
BBR 3571	0.054	0.081	0.25	261	0.25	78
BBR 3005	3.03	2.4	4.5	144	3	72
BBR 3464	0.0094	0.0075	0.25	239, 389	0.3	73
Cisplatin[b]	2.4	22.3	6	110 (100–122)	4	38

[a]See Figure 2 for structures. Data adapted from references 10 and 23. All compounds were dissolved in NaCl 0.9% and further diluted in complete culture medium. IC_{50} is Inhibiting Concentration 50% of cellular growth after 48 h of drug exposure. [b]Mean of 198 experiments. [c]10^5 cells i.p. into CD2F1 male mice. Treatment i.p. on days 1, 5, 9 after tumor transplantation (day 0). All compounds were dissolved in saline just before use. T/C: median survival time of treated mice/median survival time of controls × 100.

Table 5 *Effects of bifunctional DNA binding of mono-, di- and tri-nuclear platinum complexes*[a]

Complex	Unwinding	Bending	Interstrand	Intrastrand	$B \rightarrow Z$	HMG[b]
Cisplatin	11°	30–35°	< 5%	d(GG), d(AG) d(GNG, minor)	No	1.0
Transplatin	9°	Hinge joint	< 5%	d(GNG) major	No	No
1,1/t,t[c]	10°	Not directed	52%	d(GG) observed	Yes	0.3
BBR3464[d]	14°	—	40–50%	—	Yes	Not detached

[a]Adapted from *Biochemistry*, 1995, **34**, 15 480 and 1996, **35**, 16 705. [b]HMG recognition for global array of adducts; see also *Anti-Cancer Drug Design*, 1994, **9**, 389 and references therein. [c]$[\{PtCl(NH_3)_2\}_2H_2N(CH_2)_nNH_2]^{2+}$. [d]Data adapted from reference 24.

cross-links, a lack of rigid 'directed' bending and the ability to irreversibly form Z-DNA in appropriate purine-pyrimidine sequences. Chain length and geometry affects the kinetics of DNA binding, the cross-link preference (intra *versus* inter) and the structures of the cross-links formed. The kinetics of DNA binding of charged platinum compounds is significantly greater than for cisplatin. For a 2 h incubation (20 μM compound) on Calf Thymus DNA the relative amounts of Pt bound are 7.3: 1.8: 1 for BBR3464 (4 + charge), BBR3005 (2 +) and cisplatin respectively (even after allowing for the 3: 2: 1 Pt ratio in the compounds). The rapid binding of BBR3464 could affect sequence specificity – the high charge could lead to initial electrostatic interactions very different to small molecules such as cisplatin and the alkylating agents, leading to enhanced sequence specificity.

At this stage, long-range interstrand cross-linking is the dominant DNA-binding feature which distinguishes our compounds from *cis*-DDP, its analogues and even the alkylating agents. Using denaturing agarose gels, DNA cross-linking can be quantitated as the amount of the DNA species migrating more slowly than the single-stranded DNA. Treatment of supercoiled DNA of various superhelical density, relaxed and linearized DNAs with the platinum complexes allows a quantitative assessment of the number of DNA interstrand cross-links formed per DNA adduct. Modelling and distance considerations indicate that from 1,2 to at least 1,6 cross-links are possible for BBR3464 – from the crystal structure the distance between the two Pt–Cl centers is 23.6 Å. In a 1,6 cross-link the guanines are separated by four base pairs whereas a 1,2 cross-link is formed from guanines on neighboring base pairs. An interstrand cross-link is a relatively minor event for *cis*-DDP but there is still controversy as to the biological implications of the lesion – the adduct bends DNA and is recognized by HMG proteins.[12]

The (Pt,Pt) interstrand cross-link is implicated in cytotoxicity and the ability to irreversibly induce the $B \rightarrow Z$ conformational change is also a feature of the most potent agents.[13] Short-chain ($n = 2, 3$; ethylenediamine, propanediamine) linkers induce the $B \rightarrow Z$ transition but are very poorly cytotoxic.[14] This apparent dichotomy can be resolved because the short-chain compounds are very weak interstrand cross-linking agents. Likewise, a very relevant finding is that the presence of planar ligands in the coordination sphere (*i.e.* [*trans*-{PtCl-(pyridine)$_2$}$_2$H$_2$N(CH$_2$)$_n$NH$_2$]$^{2+}$ or   [{*trans*-PtCl(NH$_3$)(quinoline)}$_2$H$_2$N-(CH$_2$)$_n$NH$_2$]$^{2+}$ stabilizes poly(dG-dC).poly(dG-dC) in the B form.[5,15] Both compounds are efficient interstrand cross-linking agents but not potent cytotoxic agents (ID$_{50}$ in L1210 leukemia $> 15\,\mu$M in both cases). From the DNA binding view (assuming reasonably similar uptake) it is therefore not induction of a left-handed form *per se* which is cytotoxic but both the efficiency of induction and irreversibility through cross-linking contribute to cytotoxicity. DNA sequences which have the potential to form Z-DNA have been found within the mammalian cell genome including transcriptional regulatory regions[16] and DNA replication origins.[17] The induction of Z-DNA within the cell would have serious consequences with regard to transcription and DNA replication.[18]

Cooperative Effects in the Solution Structures of Site-specific (Pt,Pt) Interstrand Cross-links

A logical place, therefore, to begin to examine for the structural origin of differential effects between mononuclear and polynuclear platinum complexes is in the structures of interstrand cross-links. Interstrand cross-linking of poly(dG-dC).poly(dG-dC) by *cis*-DDP does not cause a $B \rightarrow Z$ transition.[19] The interstrand cross-link of cisplatin is a 1,2-cross-link between adjacent guanines. As stated, a very interesting NMR structure of a 1,2-interstrand cross-link in 5'-d(CATAG*CTATG)-d(GTATCG*ATAC) shows a left-handed but highly localized *Z*-DNA formation at the binding site due in part to an induced *syn* conformation of the platinated bases.[12] This conformation results in a large bending angle and the interstrand cross-link is thus recognized by HMG proteins.

The potential for systematic alteration of DNA structure by platination is demonstrated by the novel structure of a site-specific 1,2-interstrand cross-link formed by reaction of $[\{trans\text{-}PtCl(NH_3)_2\}_2H_2N(CH_2)_4NH_2]^{2+}$ with the central d(GC) pair of the symmetrical $\{5'\text{-}CATG*CATG\text{-}3'\}_2$.[20] The structure is totally different from that of the cisplatin cross-link and is a dumbbell shape caused by looping of each single complementary strand back on itself with the two 'looped' single strands stacked together by the nature of the dinuclear Pt–G binding.

For dinuclear and trinuclear complexes separated by long-chain diamines, 'nearest neighbour' 1,2 cross-linking will not be favoured over longer-range cross-links. The final structures and conformational change will also be affected by interactions of the linking backbone with the non-platinated base pairs. This combination of interactions could have the ability to induce the conformational change in a cooperative fashion beyond the actual site of platination. This is especially attractive for formation in *Z*-like sequences. The salt induced $B \rightarrow Z$ transition is highly cooperative – a four base-pair section is required to 'nucleate' adjacent sequences toward the left-handed form.[21] A four to six base-pair sequence is inherently achievable by long-range interstrand cross-links and, if altered to a left-handed form, is capable of being propagated through appropriate joining sequences. Thus, the possibility exists that the Pt–DNA adduct is disseminated through a longer piece of the helix, unlike the relatively localized distortions induced by cisplatin–DNA adducts.[22] In design of DNA-binding drugs there is much discussion of the need to target 10–15 base-pair sequences to begin to approach a gene-selective strategy. The ability to cooperatively alter conformation over large distances is an exciting and important direction to pursue and a further unique advantage of the polynuclear approach.

6   Summary

This contribution succinctly reviews the chemical and biological data of a new structural class of anticancer agents based on a poly(di, tri)nuclear structure. The differences and similarities we will observe between these agents and the 'classical' cisplatin compounds as they hopefully proceed through the clinic will add to our understanding of drug discovery and design for further rational approaches to the drug treatment of cancer.

Acknowledgements

It is a great pleasure for the authors to acknowledge the dedication and enthusiasm of all the team in Boehringer Mannheim for this project. We wish to also thank V. Brabec and F. Zunino for their valuable cooperation and insight.

References

1 M.C. Christian, *Seminars Oncol.*, 1992, **19**, 720.
2 N. Farrell, in *Advances in DNA Sequence Specific Agents*, Vol. 2, ed. L.H. Hurley and J.B. Chaires, JAI Press, New Haven, 1996, 187.
3 N. Farrell, *Comments Inorg. Chem.*, 1995, **16**, 373.
4 1.N. Farrell, *Cancer Investigation*, 1993, **11**, 578.
5 N. Farrell, T.G. Appleton, Y. Qu, J.D. Roberts, A.P. Soares Fontes, K.A. Skov, P. Wu and Y. Zou, *Biochemistry*, 1995, **34**, 15 480.
6 T.G. Appleton, Y. Qu, J.D. Hoeschele and N. Farrell, *Inorg. Chem.*, 1993, **32**, 2591.
7 C. Manzotti, D. Torriani, E. Randisi, M. De Giorgi, G. Pezzoni, E. Menta, S. Spinelli, H.H. Fiebig, N. Farrell and F.C. Giuliani, *Proc. AACR*, 1997, **38**, 2080.
8 C.C. Harris, *J. Natl. Cancer Inst.*, 1996, **88**, 1442.
9 G. Pratesi, S.C. Righetti, R. Supino, D. Polizzi, C. Manzotti, F.C. Giuliani, G. Pezzoni, S. Tognella, S. Spinelli, P. Perego, N. Farrell and F. Zunino, *Br. J. Cancer*, in press.
10 H. Rauter, R. Di Domenico E. Menta, A. Oliva, Y. Qu and N. Farrell, *Inorg. Chem.*, 1997, **36**, 3919.
11 M.C. O'Sullivan and D.M. Dalrymple, *Tetrahedron Lett.*, 1995, **36**, 3451.
12 H. Huang, L. Zhu, B.R. Reid, G.P. Drobny and P.B. Hopkins, *Science*,1995, **270**, 1842.
13 P. Wu, M. Kharatishvili, Y. Qu and N. Farrell, *J. Inorg. Biochem.*, 1996, **63**, 9.
14 A. Johnson, Y. Qu, B. Van Houten and N. Farrell, *Nucleic Acids Res.*, 1992, **20**, 1697.
15 M. Kharatishvili, M. Mathieson and N. Farrell, *Inorg. Chim. Acta*, 1997, **255**, 1.
16 Y. Kohwi, *Nucleic Acids Res.*, 1989, **17**, 4493.
17 A. Bianchi, R.D. Wells, N.H. Heintz and M.S. Caddle, *J. Biol. Chem.*, 1990, **265**, 21 789.
18 A. Rich in *Proceedings of the Robert A. Welch Foundation*, Vol. 37, Houston, 1993, p. 13.
19 A. Rahmouni and M. Leng, *Biochemistry*,1987, **26**, 7229.
20 D. Yang, S. van Boom, J. Reedijk, J. van Boom, N. Farrell and A.H.-J. Wang, *Nature Struct. Biol.*, 1995, **2**, 577.
21 F.M. Pohl and T.M. Jovin, *J. Mol. Biol.*, 1972, **67**, 375.
22 P.M. Takahara, A.C. Rosenzweig, C.A. Frederick and S.J. Lippard, *Nature*, 1995, **377**, 649.
23 H. Rauter, R. Di Domenico, E. Menta, G. Da Re, G. De Cillis, M. Conti, A. Lotto, F. Pavesi, S. Spinelli, C. Manzotti, L. Piazzoni and N. Farrell, *Proc. AACR*, 1998, **39**, 1096.
24 N. Farrell, Y. Qu, J. Kasparkova, V. Brabec, M. Valsecchi, E. Menta, R. DiDomenico, M. Conti, G. Da Re, A. Lotto and S. Spinelli, *Proc. AACR*, 1997, **38**, 2077.

CHAPTER 9

Oxidation Damage by Bleomycin, Adriamycin and Other Cytotoxic Agents That Require Iron or Copper

DAVID H. PETERING, JUN XIAO, SREEDEVI NYAYAPATI,
PATRICIA FULMER AND WILLIAM E. ANTHOLINE

Department of Chemistry, University of Wisconsin-Milwaukee and
Department of Radiation Biophysics, Medical College of Wisconsin, USA

1 Introduction

There are now more than forty drugs approved for the treatment of human cancer in the United States.* Considering the thousands of compounds that have been tested as candidate antitumor agents, this is a highly select group. Among them are two drugs that require a metal ion as part of their structures. One is the simple metal complex *cis*-diamminedichloroplatinum(II). The other, bleomycin, is a natural product that must form an iron complex to display cytotoxicity.[1] In addition, two anthracycline natural products, doxorubicin or adriamycin, and daunomycin, may also function as iron complexes or utilize cellular iron in an indirect way in their mechanisms of action.

Bleomycin causes cellular DNA damage by an oxidative mechanism; adriamycin and daunomycin also produce oxidative damage as part of their repertoire of reactions in cells. Other quinone antitumor drugs used in the clinic, including mitomycin C, mitroxanthone, and mithramycin, may also require cellular iron to catalyze formation of reactive oxidants. Thus, quite apart from

*Abbreviations: Adr, adriamycin or doxorubicin; Blm, bleomycin; BPS, 4,7-phenylsulfonyl-1.10-phenanthroline; CuKTS, 3-ethoxy-2-oxobutyraldehyde bis(thiosemicarbazonato)copper(II); GSH, glutathione; ICRF-187, (+)-1,2-bis(3,5-dioxopiperazinyl-1-yl)propane (this compound is now called ADR-925 by its manufacturer); ICRF-198, the hydrolyzed form of ICRF-187 (called ADR-529 by the manufacturer); NC, 2,9-dimethyl-1,10-phenanthroline; Phen, 1,10-phenanthroline; RSH, thiol compound; Topo II, topoisomerase II; Tr, transferrin.

deliberate interest in metallodrugs or metal-based redox chemistry, as many as 15% of the anticancer drugs approved for human use may need a metal ion for activity.

Experimental antitumor agents such as streptonigrin, bisthiosemicarbazones, and perhaps monothiosemicarbazones must form iron or copper complexes to become biologically active as catalysts of oxidant damage to cells.[2,3] In sum, metal-based catalysis of redox reactions in cells describes a major topic in cancer therapeutics. Nevertheless, it has remained underdeveloped as a theme for study and application. The sections below provide a coordinated review and perspective on the subject of redox-active, metal-dependent drugs in cancer chemotherapy.

2   Bleomycin

Bleomycin (Blm) is a natural product that is in regular use in combination chemotherapy.[4–6] The left side of its structure contains a versatile metal-binding domain, probably composed of five labeled nitrogens (Figure 1). The right side includes a bithiazole moiety and a positively charged, variable R group, which can bind to DNA. A well-defined role for the disaccharide unit has yet to be established.

As a subject for mechanistic study, Blm is an enormously important molecule. It is an effective antitumor agent and a model for an effective metallodrug. Furthermore, it is a highly efficient reagent for the generation of DNA double strand breaks. As such, it has also become the prototype for the design of synthetic chemical nucleases involving redox-active metal sites.[7]

Clinical Use

Bleomycin is commonly employed in the treatment of some human cancers.[8] It is administered as the metal-free natural product; yet, as discussed below, Blm must be converted into FeBlm to inhibit cell proliferation and degrade cellular DNA.

The relationship between the efficacy of Blm chemotherapy and its access to circulating or cellular iron has not been examined. Cancer patients frequently display a generalized stress response or cachexia which inhibits appetite. In this state, plasma iron can be markedly depressed.[9] Whether this influences the availability of iron for the drug is unknown; indeed, according to experiments described below, it appears that Blm retains full activity in iron-deficient animals.[10] However, FeBlm seems to cause less general host toxicity than Blm does.[11] Therefore, since FeBlm is the active form of the drug, it is surprising that trials have not been conducted to determine whether FeBlm offers therapeutic advantages over metal-free Blm. At the least, delivery of FeBlm would circumvent problems that the drug may have in finding sources of iron.

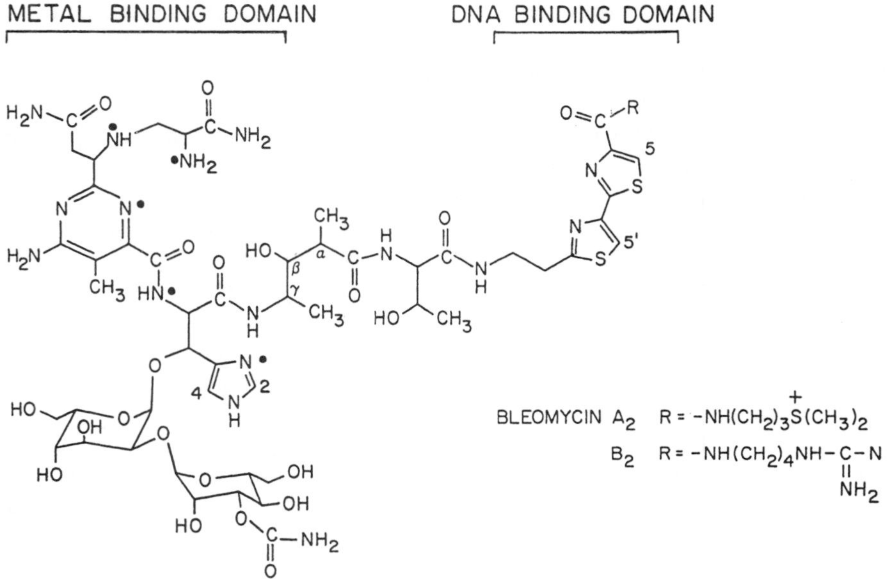

METAL BINDING DOMAIN   DNA BINDING DOMAIN

Figure 1 *Structure of bleomycin*

Animal and Cellular Studies on the Role of Iron in the Cytotoxicity of Bleomycin

There are few reports that have dealt with the interaction of Blm with biologically essential transition metal ions that may be involved in its mechanism of action. Initial reports of the isolation of Blm mentioned that the product isolated from *Streptomyces verticillus* contained significant amounts of Cu, to the extent that the powdered drug was bluish.[12] Indeed, Blm fully saturated with Cu^{2+} displayed antitumour properties similar to the original preparations of the drug.[13] Although it was originally assumed that CuBlm was the active form of the drug, metal-free Blm was as effective and caused fewer side-effects. Thus, if a metal is required for activity, it can be garnered from the host.

Soon, it was observed that the drug caused the strand cleavage of DNA and that the reaction was enhanced by the presence of Fe^{2+} and reductants.[14,15] Animal experiments were then undertaken to determine whether Fe or some other transition metal was required to elicit the activity of Blm. It was anticipated that host Fe or Cu deficiency might depress the activity of Blm, because less of the activating metal ion would be available in the organism for complexation by the drug. However, Blm retained full activity in iron- and copper-deficient mice.[10]

Early cellular experiments were unable to define FeBlm as the active form of the drug.[10] Nevertheless, two sufficiently iron-deficient cells were finally produced, which showed that cellular iron was required for both single and double

DNA strand scission and for inhibition of cell proliferation and reduction in viability;[16] exposure of cells to CuBlm did not overcome the effect of iron deficiency. One such type of cell was the microorganism *Euglena gracilis*, which could be made rigorously iron-deficient upon growth in defined medium containing less than 10 nM Fe. The other was human HL-60 cells adapted to grow in defined medium on iron-pyridoxal isonicotinoylhydrazone and then cultured in the absence of this source of iron. These experiments indicated that cells must be highly Fe-deficient before any reduction in Blm activity can be detected.

A fundamental question about the mechanism of action of Blm is how it acquires iron in the host organism. Since iron is closely regulated and exclusively bound to transferrin in plasma, and is predominantly associated with ferritin and other proteins in cells, most of it may not be readily available to Blm. However, few experiments have examined whether Blm might compete for Fe bound to metalloproteins.

A second question relates to mechanisms of biological interconversions of metallobleomycins. Blm and its Fe-, Cu-, and Zn-complexes are similarly cytotoxic to cells exposed under conditions that minimize extracellular cross reaction with other metal ions, implying that each can be converted into FeBlm.[11] Yet, Cu^{2+} binds to Blm with higher affinity than does Fe^{3+}, and cellular Zn^{2+} may well be available for reaction with Blm.[5,17] CuBlm has also been detected in urine after administration of Blm.[18] Opposite conclusions have been reached on whether copper can be removed from Blm by reductive ligand substitution using thiol reagents.[19,20]

Cellular Properties of DNA Strand Scission

Both single and double strand scission can be detected in cells exposed to Blm and its Fe(III) and Cu(II) complexes, by using alkaline and neutral filter elution methods respectively.[21–23] The extent of formation of double, but not single, strand lesions parallels the drug concentration dependence for inhibition of cell proliferation.[21] While single strand scission is readily repaired, double strand breaks persist.[24] Both findings support the central role of double strand cleavage in the cytotoxic process. These DNA lesions can also be detected in isolated nuclei exposed to FeBlm, whereas Blm and CuBlm display little, if any, capacity to damage DNA in this system.[25]

Biochemical Mechanisms of DNA Damage

Characterization of DNA Damage by FeBlm

Reactions of species of Blm with DNA have been examined in much detail in model reactions. A number of properties of these reactions have been established. First, as in cells, FeBlm-dependent reduction of dioxygen leads to a combination of single and double strand cleavage, and also base release.[4–6] Evidently, each process begins with the abstraction of a C4′ hydrogen atom on a deoxyribose site along the backbone of DNA (Figure 2).[26–28] If no O_2 is present, base release

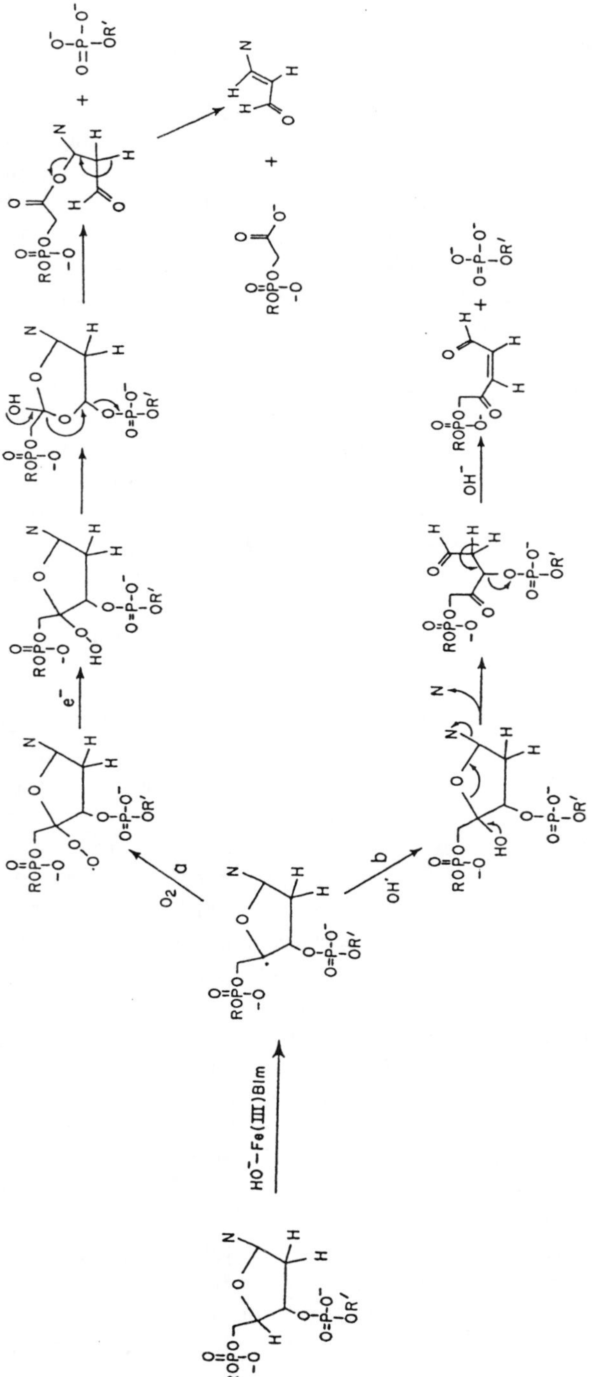

Figure 2 *Pathways of DNA damage caused by $HO_2^- $-$Fe(III)Blm$*

occurs; in the presence of O_2 the other pathway competes for the C4' radical, and frank cleavage of the backbone results.

Second, the observed cleavage patterns of supercoiled plasmids and synthetic hairpin DNA duplexes, as well as the kinetics of cleavage, demonstrate that both single and double strand cleavage occur at all concentrations of drug.[29–31] This indicates that double strand scission is not the outcome of the introduction of many single strand breaks, culminating in apparent double strand cleavage, but instead represents a subset of the DNA damage events occurring at a given site. Thus, double strand cleavage probably results from the action of a single molecule of FeBlm acting on both strands of DNA at a particular site.[31–34] Certainly, in the cell, where the concentration of drug in the nucleus is no more than one molecule per 10^5 bases, this must be the case.[21]

There is site selectivity in the reactions of FeBlm and DNA. The drug strongly prefers cleavage sites between 5'-GpC-3' or 5'-GpT-3', and release of C or T from these sites.[5,35–37] Furthermore, single strand breakage at a preferred site can be accompanied by single strand cleavage on the other strand at the same base pair or offset by one base pair, producing double strand scission, or can be accompanied by base release on the other strand.[32,33] Alternatively, base release at the preferred site occurs without further damage to the other strand.

Solution Properties of Metallobleomycins Related to the DNA Damage Reaction

The structure of Blm shown in Figure 1 may be divided into a metal-binding domain, a DNA-binding domain, a peptide linker between them and a disaccharide, which has not been the subject of much study. With the finding that Blm and ZnBlm adopt extended conformations in solution, it was assumed for a number of years that the DNA domain and the linker act to tether the reactive metal domain in the vicinity of DNA.[38,39] Then, the redox chemistry at the metal center occurs in proximity to DNA and efficiently causes damage to its structure. This description resembles that for synthetic nucleases that have been constructed using this model.[7] The nature of the interaction between the DNA domain of Blm and DNA itself has been the subject of many studies. Both the bithiazole and the variable, positively charged R group in the domain contribute to binding. Although some results have suggested an intercalative mode of binding, association along the minor groove has also been proposed.[5,6,40–43] In either case, it was originally thought that the site selectivity of the drug for DNA cleavage and base release resided in structural characteristics of the bithiazole.[6,43] As described below, this view has evolved to include the metal domain in the mechanism of site selectivity.

The early publication of the crystal structure of the metal domain of CuBlm, which lacks the disaccharide, remains the only X-ray structure of a metallobleomycin species, and so continues to serve as a model for the metal-ligand structure of various metallobleomycins.[44] As seen in Figure 3, its copper ion is bound to the five nitrogen atoms denoted by dots in Figure 1.

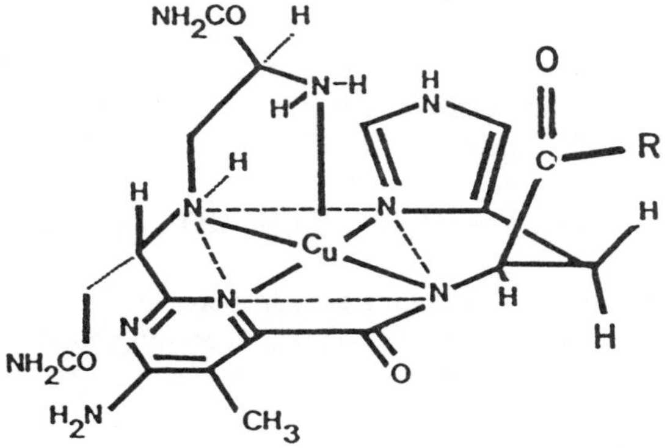

Figure 3 *Structures of CuP-3A, the modified metal domain of CuBlm*

An elegant series of papers by Peisach, Horwitz and colleagues has demonstrated that *in vitro* DNA strand scission is dependent upon the presence of iron in the reaction mixture and involves the following pathway of activation of the drug to carry out the strand scission reaction:

$$Fe^{3+} + Blm \rightarrow Fe(\text{III})Blm \tag{1}$$

$$Fe(\text{III}) + Blm + e^- \rightarrow Fe(\text{II})Blm \tag{2}$$

$$Fe(\text{II})Blm + O_2 \rightarrow O_2\text{-}Fe(\text{II})Blm \tag{3}$$

$$2O_2\text{-}Fe(\text{II})Blm + H^+ \rightarrow HO_2{}^-\text{-}Fe(\text{III})Blm + Fe(\text{III})Blm + O_2 \tag{4}$$

The reducing agents in Reaction 2 can be reagents such as thiols or ascorbate.[14] Upon addition of O_2 to Fe(II)Blm, a fleeting intermediate is observed, probably the dioxygenated species, which rapidly reacts to generate the products of Reaction 4.[45,46] That $HO_2{}^-$-Fe(III)Blm is formed in this reaction is supported by ESR, Mössbauer, mass spectroscopic, and EXAFS analysis of its structure.[46-49] It can also be formed by the direct reaction of H_2O_2 with Fe(II)Blm:[46]

$$H_2O_2 + Fe(\text{III})Blm \rightarrow HO_2{}^-\text{-}Fe(\text{III})Blm + H^- \tag{5}$$

Using Reaction 5 to generate $HO_2{}^-$-Fe(III)Blm, it has been shown that this species is competent to initiate strand scission or, at least, is the only observable precursor of a form that starts the process of DNA degradation.[46] Indeed, if DNA is not present, this species begins to attack its own structure in a reaction that inactivates it for subsequent reactions with DNA.[46,50,51] Considering the various outcomes of the reaction of $HO_2{}^-$-Fe(III)Blm with DNA as overall redox reactions (Figure 2), base release involves a simple stoichiometry of reaction:

$$HO_2{}^-\text{-}Fe(\text{III})Blm + C4'\text{-}H \rightarrow C4'\text{-}H^+ + Fe(\text{III})Blm + H_2O + OH^- \tag{6}$$

Reaction of $HO_2{}^-$-Fe(III)Blm with DNA to produce a single strand break had been conceived of in two ways: in the first, abstraction of the C4' hydrogen formally involves a species equivalent to a hydroxyl radical.[5,31] As such, there is homolytic cleavage of the peroxide, leaving a ferryl species, equivalent to a hydroxyl radical and Fe(III)Blm, as the other product.

$$HO_2{}^--Fe(III)Blm + C4'-H \rightarrow C4' \cdot + HO-Fe(IV)Blm + H_2O \qquad (7)$$

As this C4' radical is converted into a hydroperoxide, an electron is required from some source in the reaction mixture. According to the other proposal, $HO_2{}^-$-Fe(III)Blm undergoes heterolytic cleavage to produce hydroxide and a perferryl species that does the hydrogen abstraction.[34] The net reaction is the same as shown in Reaction 7:

$$HO_2{}^--Fe(III)Blm \rightarrow O=Fe(V)Blm + OH^- \qquad (8)$$

$$O=Fe(V)Blm + C4'-H \rightarrow HO-Fe(IV)Blm + C4' \cdot \qquad (9)$$

It is further assumed that the other electron needed to reach the HO_2-C4' intermediate is supplied by the ferryl product in Reaction 10:

$$HO-Fe(IV)Blm + C4' \cdot + O_2 + H^+ \rightarrow HO_2\text{-}C4' + O=Fe(V)Blm \qquad (10)$$

In this model, the perferryl species is one of the products. Thus, in neither hypothetical pathway of single strand scission is Fe(III)Blm generated. Instead, a reactive species remains. How either ferryl or perferryl iron centers are reduced to Fe(III) is unresolved.

The problem of the extra oxidant species in isolated single strand cleavage is an asset for the construction of mechanisms of double strand cleavage by single molecules of FeBlm. Although it is unknown whether Fe(III)Blm must be activated twice to obtain double strand cleavage, in principle, either of the mechanisms above (Reaction 7 or Reactions 8–10) can provide a reactive FeBlm species to abstract a hydrogen atom from the C4'-H on the second strand:

$$HO-Fe(IV)Blm + C4'-H \rightarrow C4' \cdot + Fe(III)Blm + H_2O \quad (11) \text{ (ref. 31)}$$

or

$$O=Fe(V)Blm + C4'-H \rightarrow C4' \cdot + HO-Fe(IV)Blm \qquad (12) \text{ (ref. 34)}$$

In the second pathway, $O=Fe(V)Blm$ will again be generated as HO_2-C4' is formed. This raises the interesting question of whether $HO_2{}^-$-Fe(III)Blm might serve as a catalytic reagent for double strand cleavage.[34] Evidence to date indicates that strand cleavage is stoichiometrically dependent upon added reducing equivalents or added $HO_2{}^-$-Fe(III)Blm.[34,52]

When single strand scission on the site-specific strand is followed by base release on the second strand, the presence of HO-Fe(IV)Blm at the site will provide only one of the two oxidizing equivalents needed to start this pathway (Reaction 6). In contrast, O=Fe(V)Blm can supply both equivalents. Thus, neither of these formulations for the reaction of a single molecule of $HO_2{}^-$-Fe(III)Blm with a DNA site is completely satisfactory.

Properties of Metallobleomycins Bound to DNA

Early studies of the interaction of ON-Fe(II)Blm and OC-Fe(II)Blm with DNA showed that DNA perturbs the metal site. The ESR signal of the former was dramatically changed by DNA, and the chemical shift of the NMR signal of the H2 proton of metal-bound imidazole in the latter was altered upon binding to DNA.[53,54] Furthermore, the low- to high-spin transition of Fe(III)Blm at pH 4, thought to result from protonation of the axial amine in the metal domain, was reversed upon addition of DNA, suggesting substantial interaction between DNA and the iron coordination site.[55] Thus, it is evident that the metal domain chemistry of FeBlm is perturbed by the presence of DNA.

What has become clear recently is that the alteration of metal domain properties upon binding to DNA occurs only when the structure interacts with specific 5'-GpPyrimidine-3' sites.[56] Such evidence along with NMR structural results described below strongly link the metal domain structure with site specification that precedes the DNA damage reactions.

In order to systematically investigate the roles that DNA may play in the chemistry of FeBlm bound to DNA, the properties of CoBlmDNA were first examined. The cobalt substitution was made because the dioxygen chemistries of Fe(II) and Co(II) are similar, but the redox reactions of CoBlm are slower than those of FeBlm and are thus amenable to more detailed study. The pathway of reaction of Co(II)Blm with O_2 is analogous to that proposed for Fe(II)Blm:[57]

$$Co(II)Blm + O_2 \rightarrow O_2\text{-}Co(II)Blm \qquad (13)$$

$$2O_2\text{-}Co(II)Blm \rightarrow BlmCo(II)\text{-}O_2\text{-}Co(II)Blm + O_2 \qquad (14)$$

$$BlmCo(II)\text{-}O_2\text{-}Co(II)Blm + H^+ \rightarrow HO_2{}^-\text{-}Co(III)Blm + Co(III)Blm \qquad (15)$$
$$\text{(Form I)} \qquad \text{(Form II)}$$

In this case, however, the presence of each species and the kinetics of reaction of each step can be observed, so that the detailed reaction mechanism above can be set forth. The products are analogous to those formed starting with Fe(II)Blm. In contrast, they are stable and can be conveniently studied.

The rate of product formation is markedly changed in the presence of DNA.[58] As the ratio of base pairs to CoBlm increases and the average distance between drug molecules distributed along the DNA structure lengthens, the rate of Reaction 14 plunges, because metal domains cannot dimerize or make contact to accomplish electron transfer. As a result, O_2-Co(II)BlmDNA becomes highly stable at base pair to drug ratios of 10 or more.

When the reaction of Fe(II)Blm with O_2 was examined in the presence of DNA, a similar mechanistic pathway was defined.[58] With ratios of base pairs to drug of 20 or greater, O_2-Fe(II)BlmDNA is stabilized, indicating that Reaction 4 is no longer a feasible reaction to generate the $HO_2{}^-$-Fe(III)Blm needed to initiate the DNA damage pathways. Interestingly, O_2-Co(II)BlmDNA and O_2-Fe(II)BlmDNA are unusually resistant to dissociation of O_2, implying that the dioxygen complex is stabilized by DNA.[59,60] Given the fact that in cells DNA is in overwhelming excess relative to nuclear Blm ($\sim 10^5$ to 1), it is evident that Reaction 4 cannot contribute to the activation process.[21] Thus another source of electrons must be involved in this step as well as in Reaction 2.

$$O_2\text{-Fe(II)BlmDNA} + e^- + H^+ \rightarrow HO_2{}^-\text{-Fe(III)BlmDNA} \qquad (16)$$

The stabilization of dioxygenated Co(II)- and Fe(II)-Blm on DNA suggests that the complex is not able to undergo the rapid cycles of dissociation and rebinding necessary to find a partner to carry out Reactions 4 or 14. Nor is a rapid sliding process available to facilitate these reactions.

Structural Properties of CoBlm Species in Solution and Bound to DNA

The conformational relationship between the metal domain of several metallobleomycins and DNA has been probed by ESR studies of Fe(III)Blm, NO-Fe(II)Blm, and O_2-Co(II)Blm bound to oriented fibers of DNA.[61,62] Remarkably, the dioxygen ligand of O_2-Co(II)Blm and the nitrosyl group of NO-Fe(II)Blm, which largely bear the unpaired spins in these complexes, are rigorously constrained to a plane approximately perpendicular to the helix axis. These results provide clear evidence that the metal domain in these adducts, like the DNA domain, must be closely associated with DNA, and not loosely tethered to the polymer. They also indicate that Fe- and Co-Blm species bind to DNA with similar steric constraints.

The stability of $HO_2{}^-$-Co(III)Blm (Form I) as well as its diamagnetic character offer the opportunity to determine its structure and compare it with that of Co(III)Blm (Form II). A two-dimensional NMR study has revealed that both of these molecules share similar, highly folded metal domain and linker regions that form close packed structures.[63,64] They differ in that the bithiazole group of Form I is also folded back upon the metal domain, generating a globular conformation for the structure (Figure 4). Consequently, the sixth coordination site holding the peroxyl group in Form I is a well-defined pocket, and the site in Form II is more open.

Considering the interaction of the Form I structure with the DNA minor groove, where binding has been thought to occur, and assuming that $HO_2{}^-$ exists in the same conformation as O_2 in O_2-Co(II)BlmDNA, one postulates that both the DNA and metal domains will be in close contact with DNA. Furthermore, the peroxide ligand will be directed into the interior of the minor groove in

DIMETHYL-
SULFONIUM   BITHIAZOLE (B5, B5′)

LINKER

PEROXIDE (—)

COBALT (∗)

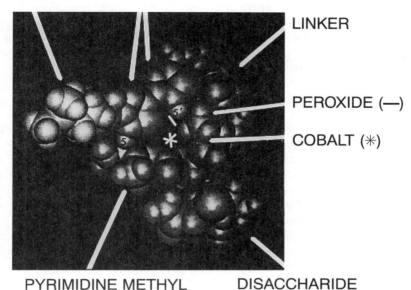

PYRIMIDINE METHYL   DISACCHARIDE

Figure 4 *Solution conformation of HO$_2$$^-$-Fe(III)Blm*

proximity to the C4′ hydrogens that are the initial target of reaction for HO$_2$$^-$-Fe(III)Blm.

To test these ideas, the structure of HO$_2$$^-$-Co(III)Blm (Form I) bound to DNA oligomers have been under study.[65–67] The key to this work was finding that both Form I and Form II bind tightly to DNA, suggesting that they interact in slow exchange on the NMR time-scale.[59] Indeed, it has been demonstrated that Form I binds in slow exchange to several oligomers, which incorporate the selective sites of strand cleavage or base release, 5′-GpC-3′ or 5′-GpT-3′, namely, d(CCAGGCCTGG)$_2$, d(CCAGTACTGG)$_2$, d(GGAAGCTTCC)$_2$, and d(AAACGTTT)$_2$.[65–67] Using the first two of these oligomers, Stubbe and co-workers showed that Form I binds to DNA in part by intercalation of the bithiazole between the base pairs C6–C7 of the first 10-mer and T5–A6 of the second oligomer.[66,67] Hydrogen bonds are made between the ring nitrogen and adjacent amine group of the pyrimidinyl ligand of Blm and N-3 and 2-amino nitrogens on the minor groove edge of **G** of the first two DNA structures above.[66,67] The result is the formation of an unusual base-triple structure in the minor groove comprised of the **G**-pyrimidine base pair and the pyrimidinyl ring of the drug. In the process, the peroxide ligand becomes oriented toward one of the two C4′ hydrogens in the minor groove as if in preparation for hydrogen atom abstraction.

For purposes of specification of the site of reaction, it is the metal domain that interacts invariantly with guanine. That the metal domain participates in site specification had been hypothesized.[6,68] Intercalation appears to be nonspecific in that the bithiazole-tail can intercalate between at least four pairs of stacked bases, each of which contains one of the elements of site selectivity, cytosine or thymine. Nevertheless, combined structural determinants in both metal and DNA domains, and perhaps in the linker region, would appear necessary to specify both the guanine and pyrimidine components of the cleavage sites.

Role of the Metal Domain in Site Selectivity

Papers by Hecht and co-workers have examined the role of the metal and DNA domains in the site selectivity of DNA damage caused by FeBlm. Phleomycin, which contains a partially reduced thiazole ring that is no longer planar, cleaves DNA with the same sequence specificity as does Blm.[69,70] However, a mono-thiazole derivative shows no sequence selectivity.[71] Both are less efficient agents than FeBlm, perhaps because they do not bind DNA as well as FeBlm does. Finally, the DNA domain of Blm tethered to Fe-EDTA indiscriminately cleaves DNA.[72] Thus, modifications of the DNA domain may have either no effect or a dominant effect on sequence selectivity. Interestingly, when the metal and DNA domains are bound together by oligoglycine linkers of 1 to 4 monomers, cleavage site specificity does not change.[68] Although this has been construed to mean that the metal domain determines the site of strand scission, the metal domain tethered to distamycin causes DNA breakage based on the binding preference of distamycin, not Blm. These mixed results point to a joint role of the whole folded structure in binding and site selection.

3 Adriamycin

Adriamycin has been heavily used for many years as a chemotherapeutic agent for the treatment of a variety of cancers, particularly solid tumors.[73] The adriamycin structure consists primarily of the fused-ring, anthracycline chromophore connected to an amino sugar residue (Figure 5). Through its set of quinone–hydroquinone functional groups, the drug can undergo complex oxidation–reduction chemistry.[74–76] Furthermore, the positioning of these groups offers reasonable binding sites for metal ions.[77] If Fe^{2+} or Fe^{3+} is bound, redox chemical options are added to the molecule. Finally, the planarity and extensive double bond character of the anthracycline portion of the molecule gives it both the lipid solubility to interact with membranes and the structure to intercalate between DNA base pairs.[78–80]

Because of the multiple facets to the chemistry of Adr, different sites and modes of action of the drug have been hypothesized. It can bind to DNA and may inhibit DNA replication after intercalation.[80,81] The drug may undergo redox cycling in the presence of cellular reductants and O_2 to generate oxy-radicals like the hydroxyl radical, which directly degrade DNA.[76,82]

$$Adr + e^- \rightarrow Adr\cdot \qquad (17)$$

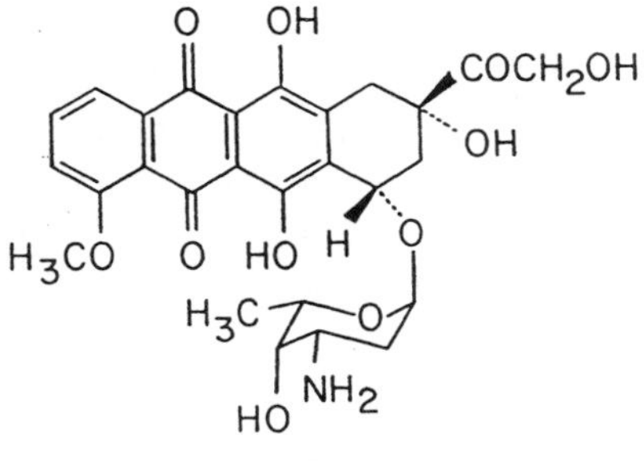

ADRIAMYCIN
(DOXORUBICIN)

Figure 5 *Structure of adriamycin*

$$Adr^-\cdot + O_2 \rightarrow Adr + O_2^-\cdot \tag{18}$$

$$2O_2^-\cdot + 2H^+ \rightarrow O_2 + H_2O_2 \tag{19}$$

$$H_2O_2 + e^- \xrightarrow{Fe/Cu} OH\cdot + OH^- \tag{20}$$

Indeed, this mechanism might include direct binding of Fe to catalyze hydroxyl radical formation.[83,84] Alternatively, the lipid solubility of Adr may direct it to the cellular membrane, where its capacity to undergo redox reactions and bind Fe may lead to lipid peroxidation and destruction of membrane analogous to its proposed effects on DNA.[85,86]

A number of proposed mechanisms of action of Adr postulate direct or indirect roles for redox-active metal ions. However, since Adr is administered as a metal-free preparation, any interactions between the drug and metal ions must occur after it enters the organism. Generally, attention has centered in the essential redox-active metals, copper and iron. Copper adriamycin is not stable in plasma and thus is unlikely to display significant stability in biological systems.[87] A complex of Fe(II) and Adr called quelamycin received some attention years ago, but its ill-defined nature weakened interest in biological Fe-Adr interactions.[88] Nevertheless, chemically rigorous studies of iron-adriamycin chemistry and biochemistry have been undertaken, and those are considered below. In addition, for those exploring the elaborate redox chemistry of Adr and its participation in the drug's cytotoxic activities, a question of wide concern has been the role of biological sources of iron as catalysts of the Fenton reaction,

$$Fe(III) + e^- \rightarrow Fe(II) \tag{21}$$

$$H_2O_2 + Fe(II) \rightarrow Fe(III) + OH\cdot + OH^- \tag{22}$$

which could participate in the reductive activation of O_2 to the hydroxyl radical by Adr (Reactions 17–22).

Besides proposed mechanisms that focus on oxidant-based damage to cells, the other hypothesized pathway of cytotoxicity with extensive support involves the inhibition of topoisomerase II (Topo II) by adriamycin.[89,90] It is hypothesized that the inhibition of Topo II as it participates in DNA replication causes DNA double strand breakage. The nature of the interaction between Adr and Topo II has not been determined, nor has a link between cellular iron and inhibition of this enzyme been suggested.

Aspects of Clinical Use

The application of Adr in cancer treatments is limited by severe, life-threatening cardiomyopathy (heart muscle degeneration), which occurs in humans and animals after a certain cumulative dose of the drug.[91–93] Building on the ideas developed above, that Adr exerts its effect through the generation of reactive O_2 species, it has been argued that heart muscle, with its intense oxygen metabolism and relatively low concentration of antioxidants such as superoxide dismutase and catalase, should be a favorable target for oxidant damage by Adr.[93,94]

Efforts have been ongoing to find means of preventing the cardiotoxicity of Adr. Given the centrality of iron in the mechanisms of redox-based toxicity of the drug, one approach has been to try to deny Adr access to reactive pools of cellular Fe that catalyze Reactions 20–22. Attention has focused on the proligand, (+)-1,2-bis(3,5-dioxopiperazin-1-yl)propane, called ICRF-198 (ADR-529), which is easily hydrolyzed to a tetradentate ligand, ICRF-187 (ADR-925), containing two carboxyl and two amino groups that can bind a number of transition metal ions (Figure 6). Administration of ICRF-187 to animals and humans effectively ameliorates the cardiotoxicity of Adr.[95,96] ICRF-198 can both compete successfully with Adr for Fe(II) and remove Fe from transferrin.[97,98]

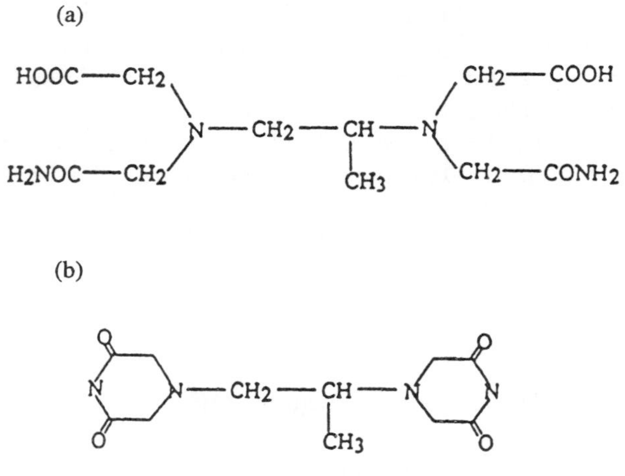

(a)

(b)

Figure 6 *Structures of* (a) *ICRF-187 and* (b) *ICRF-198*

Cytotoxic Mechanisms and the Role of Iron

Despite the many studies that have been conducted on the mechanism of cytotoxic action of Adr in tumor and heart, and those that have documented interactions between ICRF-187 and Adr, a role for iron in these processes has yet to be defined. In an effort to determine whether iron is required for the cytotoxic actions of Adr, several types of cells have been made iron deficient and tested for sensitivity to Adr.[99,100]

Adriamycin is completely without effect on cell proliferation in severely iron-deficient *Euglena gracilis* and has substantially diminished activity in the Fe-deficient mammalian HL-60 tumor and H9c2 (2-1) heart myoblast cell lines.[99,100] Thus, the availability of cellular iron is necessary for much, if not all, of the cytotoxic effects of Adr in tumor and cardiac cell models.

Biochemical Mechanisms of Reaction of Adriamycin with Iron

One class of mechanisms for the biological activity of Adr assumes that $FeAdr_n$ ($n = 2$ or 3) can form and exist long enough to target sensitive sites in the cell. It may bind to DNA through its anthraquinone ring system and positively charged amino sugar, and at the same time move Fe necessary for the production of hydroxyl radicals into close proximity to DNA.[84] Alternatively, the iron complex may associate with membrane and initiate lipid peroxidation.[93,101,102] Whether or not $Fe(III)Adr_n$ has the stability to exist in the cell or in its surrounding medium, where a plethora of other iron-binding ligands also exist, has yet to be decided.[103,104] If it is unstable, then the indirect mechanisms of interaction between Adr and pools of cellular iron gain in interest.

Model studies have suggested that Adr may be able to mobilize iron from physiological iron binding sites such as transferrin and ferritin through either reductive dissociation or direct binding of $Fe(III)$.[105–108] The reactions of Adr and reducing agents with ferritin or microsomes to liberate iron were conducted in the presence of the $Fe(II)$ binding ligand, 1,10-phenanthroline; those with transferrin occur only at pH 6 and below, thereby excluding reaction with transferrin circulating in the blood.[102,106–108] In the first two cases, Adr serves as an oxidation–reduction site to deliver electrons to reduce $Fe(III)$ associated with ferritin or microsomal membranes to $Fe(II)$, which is more labile and available for chelation. The nonphysiological acceptor ligand 1,10-phenanthroline may be required as a thermodynamic sink for $Fe(II)$ to drive these reactions.[102,106,107] For the reduction of Adr to Adr$^-\cdot$, a variety of one electron donors have been employed, including NADPH cytochrome P-450 reductase and NADH dehydrogenase.[106,109]

Once $Fe(III)Adr_n$ is formed, it slowly undergoes internal oxidation–reduction,

$$Fe(III)Adr_n(ROH) \rightarrow Fe(II)Adr_n(R\!=\!O)^- \cdot \qquad (23)$$

in which $Fe(III)$ is reduced, a quinone function is converted into its semiquinone and its side chain alcohol is oxidized.[101,110] In the presence of dioxygen, the

Fe(III)-anthracycline is regenerated, and the hydroxyl radical is formed, presumably as an end product of the initial formation of superoxide anion.[111] Fe(III)Adr$_n$ can also react with glutathione to yield Fe(II)Adr$_n$, which, in turn, can transfer an electron to dioxygen, or can generate a hydroxyl radical with hydrogen peroxide.[112,113] Ascorbate is also an effective reductant of Fe(III)Adr$_n$.[114]

Biological Model Reactions of Fe(III)Adr$_n$

Fe(III)Adr$_n$ causes lipid peroxidation and induces DNA damage *in vitro* in the presence of reductants.[111-115] In cells, the cleavage of DNA by Adr is inhibited under iron-deficient conditions, but is insensitive to the presence of added catalase.[116] In contrast, Fe(III)Adr$_n$ causes DNA strand scission that is largely eliminated by the addition of catalase. The first result extends the iron-dependence of cytotoxicity described above to a particular site of reaction. The second shows that, at least when Fe(III)Adr$_n$ is used, DNA strand scission results from a cascade of reactions leading to oxyradical formation, which can be interrupted by the dismutation of H_2O_2 to O_2 and H_2O by catalase.

4   Redox Active Copper Complexes and the Design of New Drugs

To design new metallodrugs that cause oxidant damage, several basic considerations have to be addressed.[3] The proposed complex must have a relatively large stability constant so that it can exist in the midst of an organism full of compounds with the potential to chelate metal ions. If the metal-free ligand is administered and metal complexation is to occur *in vivo*, the ligand must have the capacity to compete for the limited intracellular pools of iron or copper. The parent ligand or metal complex must be able to diffuse or be transported across cell membranes to reach target sites of reaction. Once in cells, the redox chemistry of the drug has to be accessible to cellular reductants such as ascorbate or glutathione. Ideally, features of the structure of the drug will direct it toward a specific site of localization and reaction, such as DNA. Examples of these properties are described below.

Mono- and bis(thiosemicarbazones) were among the first compounds that were deliberately constructed as metal-binding ligands for the purpose of inhibiting and killing tumor cells (Figure 7).[117,118] They have formed the basis for continuing inquiry into the mechanisms of action of cytotoxic copper complexes. In turn, these studies reveal the breadth of reactivity that can be selectively modulated to design new chemotherapeutic agents.

Bis(thiosemicarbazones) have been made that display powerful activity against established solid tumours in rodents, and that are dependent upon the presence of copper in the host.[119] The N_2S_2 metal coordination site in these ligands reacts with Cu^{2+} to form highly stable complexes. Thus, ligand substitution reactions are unfavorable. As the number of alkyl groups increases, the redox potential declines as much as 100–200 mV, until it is out of range for

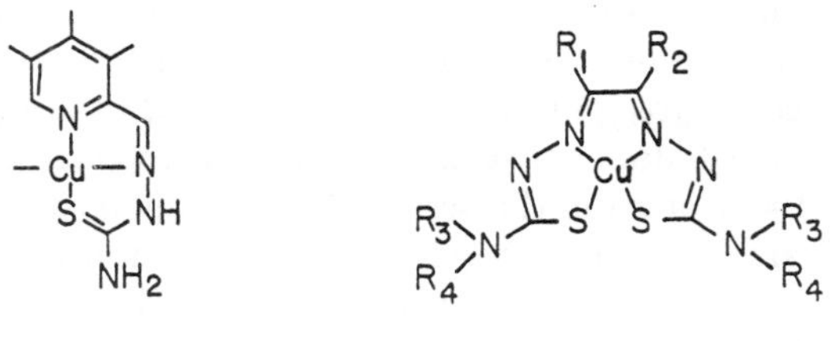

	R_1	R_2	R_3, R_4
CuKTS	$CH_3CH_2OCHCH_3$	H	H
CuKTSM$_2$	$CH_3CH_2OCHCH_3$	H	CH_3

Figure 7 *Structures of thiosemicarbazonato Cu(II) complexes: (left) monothiosemicarbazones; (right) bis(thiosemicarbazones) CuKTS and CuKTSM$_2$*

reaction of the N^4-dimethyl species with ascorbate, and is at the lower end of the range of facile reduction by glutathione ($E^{o\prime} = -240\,mV$).[120] This trend is directly reflected in the rates of reductive dissociation of these copper complexes in Ehrlich tumor cells, such that CuKTS is reductively dissociated 250 times faster than is CuKTSM$_2$.[121] In turn, R group substitution is directly correlated with antitumor and cytotoxic effectiveness.[121] Synthesizing this information, the initial reaction between Cu(II)KTS and cells is thought to be the following:

$$Cu(II)KTS + RSH \rightarrow Cu(I)KTS + RS\cdot\ (\tfrac{1}{2}RSSR) + H^+ \qquad (24)$$

$$Cu(I)KTS + RSH + H^+ \rightarrow Cu(I)SR + H_2KTS \qquad (25)$$

Once Cu(I)SR is formed, Cu has the capacity to couple the oxidation–reduction reaction of thiols or other reductants with O_2. In particular, one can envision the following sequence of reactions:

$$Cu(I)SR + O_2 \rightarrow Cu(II)SR + O_2^{-}\cdot \qquad (26)$$

$$Cu(II)SR \rightarrow Cu(I) + \tfrac{1}{2}RSSR \qquad (27)$$

$$2O_2^{-}\cdot + 2H^+ \rightarrow O_2 + H_2O_2 \qquad (28)$$

$$Cu(I) + H_2O_2 \rightarrow Cu(II) + OH\cdot + OH^- \qquad (29)$$

Reaction 28 may occur spontaneously or be catalyzed by superoxide dismutase or a copper complex. The production of the hydroxyl radical may be responsible for the cytotoxic properties of CuKTS. Alternatively, the oxidation of sensitive protein thiols may also elicit cytotoxicity.

Copper bis(thiosemicarbazones) are neutral complexes that also display a range of partition coefficients between nonpolar solvents and water.[121] For example, CuKTS exhibits substantial solubility in both water and 1-octanol, whereas $CuKTSM_2$ prefers nonpolar media. Thus, solubility and partition coefficients are readily adjustable parameters, which should modulate both initial uptake by cells and distribution between their aqueous and membrane fractions.

The cytotoxic activity of monothiosemicarbazones such as α-N-heterocyclic aldehyde thiosemicarbazones is enhanced upon formation of iron or copper complexes.[122] Indeed, upon administration of 5-hydroxy-2-formylpyridine thiosemicarbazone to humans, its Fe(II) complex is readily detected in urine.[123] Because they involve tridentate ligands, copper monothiosemicarbazones differ from bis(thiosemicarbazonato)Cu(II) complexes in that they can form adducts with Lewis bases.[124] In various model systems including cells, thiol and histidine-like adducts of 2-formylpyridine thiosemicarbazonato Cu(II) (CuL) have been detected by ESR spectroscopy.[125,126] That alters their biological chemistry, because drug distribution becomes dependent in part upon the possibilities for adduct formation. Copper monothiosemicarbazones, like bis(thiosemicarbazonato) Cu complexes, can support reduction of O_2 to reactive, toxic species. An important difference is that they do so while maintaining a modicum of structural stability.[125,126] The following sequence of reactions

$$\text{Cu(II)L} + \text{RSH} \rightarrow \text{Cu(II)-SR} + \text{H}^+ \tag{30}$$

$$\text{Cu(II)L-SR} \rightarrow \text{Cu(I)L} + \text{RS} \cdot (\tfrac{1}{2}\text{RSSR}) \tag{31}$$

$$\text{Cu(I)L} + \text{O}_2 \rightarrow \text{Cu(II)L} + \text{O}_2^- \cdot \tag{32}$$

$$\text{Cu(I)L} + \text{RSH} \rightarrow \text{Cu(I)SR} + \text{HL} \tag{33}$$

shows that a key distinction between the pathways of reaction of mono- and bis(thiosemicarbazone) copper complexes is that direct reoxidation of Cu(I)L competes with ligand substitution to form Cu(I)SR (Reactions 32 and 33), whereas reoxidation is apparently not an important pathway for bis(thiosemicarbazonato) Cu(I). In principle, if the ligand structure can be designed to favor association with particular sites within the cell, then site-specific, catalytic generation of hydroxyl radicals can be maintained with this type of copper complex.

Recent studies with copper phenanthroline complexes expand on the possibilities for site-directed oxidation–reduction chemistry of copper complexes. $Cu(II)(1,10\text{-phenanthroline})_2$, alone or tethered to various DNA-binding domains, causes DNA strand scission *in vitro* in the presence of reductants, which involves the formation of the hydroxyl radical or its equivalent.[7,127] Upon reaction of $Cu(II)(Phen)_2$ with tumor cells, it is likely that the complex binds directly to DNA, acting as a site-directed catalyst for the generation of oxyradicals.[128]

An intriguing variation of this model involves the cellular chemistry of $Cu(II)(2,9\text{-dimethyl-1,10-phenanthroline})_2$ [$Cu(NC)_2$]. $Cu(Phen)_2$ readily catalyzes the redox reaction of reductants with O_2, because $Cu(I)(Phen)_2$ is an unstable oxidation state in the presence of O_2. In contrast, $Cu(NC)_2$ does not catalyze these reactions when a noncoordinating reductant such as ascorbate is used, because $Cu(I)(NC)_2$ is stable in aerobic solution. Nevertheless, $Cu(II)(NC)_2$, which is rapidly reduced upon uptake into tumor cells, is a highly effective redox catalyst for strand breakage of DNA.[129]

An initial hypothesis to explain these observations was that glutathione (GSH), the prevalent tripeptide thiol in cells, competes effectively for $Cu(I)$, despite the large stability constant of $Cu(NC)_2$ at pH 7.4 of approximately 10^{19}, and that it then serves as the site for catalysis of reduction of O_2 by thiols, as in Reactions 26–29.[129,130] However, in unpublished model experiments, it has been shown that catalysis by the reaction mixture of $Cu(NC)_2$ and GSH is much more effective than that by CuSG alone.[131] This has led to a modified model for reaction, in which a glutathione adduct is a key intermediate:

$$Cu(II)(NC)_2 + GSH \rightarrow Cu(II)(NC)SG + NC + H^+ \qquad (34)$$

$$Cu(II)(NC)SG \rightarrow Cu(I)(NC) + RS \cdot (\tfrac{1}{2}RSSR) \qquad (35)$$

$$Cu(I)(NC) + O_2 \rightarrow Cu(II)(NC) + O_2^{-} \cdot \qquad (36)$$

$$Cu(II)(NC) + GSH \rightarrow Cu(II)(NC)SG + H^+ \qquad (37)$$

The adduct hypothetically participates in reaction rate enhancement in two ways. First, it undergoes oxidation–reduction faster than Cu(II)SG does because the presence of NC in the coordination sphere raises the redox potential of Cu. Second, upon internal oxidation–reduction, a coordination site becomes available for rapid inner-sphere reaction with O_2.

In summary, there appears to be a wide variety of reaction pathways involving copper complexes that can cause oxidant damage in cells. These take advantage of different chemical characteristics of various complexes and may be able to target particular sites of reaction in cells.

Acknowledgements

The authors appreciate the support of NIH grant CA-22108 and American Cancer Society grant DHP 31.

References

1 G. Powis, in *The Toxicity of Anticancer Drugs*, ed. G. Powis and M.P. Hacker, Pergamon Press, New York, 1991, p. 1.

2 J. Hajdu, in *Metal Ions in Biological Systems*, ed. H. Sigel, Marcel Dekker, Basel, 1985, Volume 19, p. 53.

3 D.H. Petering and W.E. Antholine, in *Reviews in Biochemical Toxicology*, ed. E. Hodgson, J.R. Bend and R.M. Philpot, Elsevier, New York, 1988, Vol. 9, p. 225.

4 J. Stubbe and J.W. Kozarich, *Chem. Rev.*, 1987, **87**, 1107.

5 D.H. Petering, R.W. Byrnes and W.E. Antholine, *Chem.-Biol. Interact.*, 1990, **73**, 133.

6 S.A. Kane and S.M. Hecht, in *Progress in Nucleic Acid Research and Molecular Biology*, ed. W.E. Cohn and K. Moldave, Academic Press, New York, 1994, p. 313.

7   D.S. Sigman and C-h.B. Chen, in *Annual Review of Biochemistry*, ed. C.C. Richardson, J.N. Abelson, A. Meister and C.T. Walsh, Annual Reviews, Inc., Palo Alto, 1990, p. 207.

8   G. Powis, in *The Toxicity of Anticancer Drugs*, ed. G. Powis and M.P. Hacker, Pergamon Press, New York, 1991, p. 106.

9   L.L. Moldawer, M.A. Marino, H. Wei, Y. Fong, M.L. Silen, G. Kuo, K.R. Manogue, H. Vlassara, H. Cohen, A. Cerami and S.F. Lowry, *FASEB J.*, 1989, **3**, 1637.

10 S. Lyman, P. Taylor, F. Lornitzo, A. Weir, D. Stone, W.E. Antholine and D.H. Petering, *Biochem. Pharmacol.*, 1989, **38**, 4273.

11 E.A. Rao, L.A. Saryan, W.E. Antholine and D.H. Petering, *J. Med. Chem.*, 1980, **23**, 1310.

12 H. Umezawa, Y. Suhara, T. Takita and K. Maeda, *J. Antibiot.*, 1966, **19**, 210.

13 M. Ishizuka, H. Takayama, T. Takeuchi and H. Umezawa, *J. Antibiot.*, 1967, **20**, 15.

14 E.A. Sausville, J. Peisach and S.B. Horwitz, *Biochemistry*, 1978, **17**, 2740.

15 E.A. Sausville, J. Peisach and S.B. Horwitz, *Biochemistry*, 1978, **17**, 2746.

16 K. Radtke, F.A. Lornitzo, R.W. Byrnes, W.E. Antholine and D.H. Petering, *Biochem. J.*, 1994, **302**, 655.

17 D. Solaiman, E.A. Rao, W.E. Antholine and D.H. Petering, *J. Inorg. Biochem.*, 1980, **12**, 201.

18 S. Hori, T. Sawa, T. Yoshioka, T. Takita, T. Takeuchi and H. Umezawa, *J. Antibiot.*, 1974, **27**, 489.

19 J.H. Freedman, S.B. Horwitz and J. Peisach, *Biochemistry*, 1982, **21**, 2203.

20 W.E. Antholine, D. Solaiman, L.A. Saryan and D.H. Petering, *J. Inorg. Biochem.*, 1982, **17**, 75.

21 R.W. Byrnes, J. Templin, D. Sem, S. Lyman and D.H. Petering, *Cancer Res.*, 1990, **50**, 5275.

22 R.W. Byrnes and D.H. Petering, *Radiat. Res.*, 1993, **134**, 343.

23 R.W. Byrnes and D.H. Petering, *Radiat. Res.*, 1994, **137**, 162.

24 R.W. Byrnes and D.H. Petering, *Biochem. Pharmacol.*, 1991, **41**, 1241.

25 R.W. Byrnes and D.H. Petering, *Biochem. Pharmacol.*, 1994, **48**, 575.

26 J.C. Wu, J.W. Kozarich and J. Stubbe, *Biochemistry*, 1985, **24**, 7562.

27 J.W. Kozarich, L. Worth, Jr., B.L. Frank, D.F. Christner, D.E. Vanderwall and J. Stubbe, *Science*, 1989, **245**, 1396.

28 L. Worth, Jr., B.L. Frank, D.F. Christner, M.J. Absalon, J. Stubbe and J.W. Kozarich, *Biochemistry*, 1993, **32**, 2601.

29 L.F. Povirk, Y.-H. Han, W. Kühnlein and F. Hutchinson, *Nucleic Acids Res.*, 1977, **4**, 3573.

30 M.J. Absalon, J.W. Kozarich and J. Stubbe, *Biochemistry*, 1995, **34**, 2065.

31 D.H. Petering, P. Fulmer, W. Li and Q. Mao, in *Genetic Responses to Metals*, ed. B. Sarkar, Marcel Dekker, New York, 1995, pp. 185–200.

32 R.J. Steighner and L.F. Povirk, *Proc. Natl. Acad. Sci. USA*, 1990, **87**, 8350.

33 L.F. Povirk, Y.-H. Han and R.J. Steighner, *Biochemistry*, 1989, **28**, 5808.

34 M.J. Absalon, W. Wu, J.W. Kozarich and J. Stubbe, *Biochemistry*, 1995, **34**, 2076.

35 A.D. D'andrea and W.A. Haseltine, *Proc. Natl. Acad. Sci. USA*, 1978, **75**, 3608.

36 M. Takeshita, A.P. Grollman, E. Ohtsubo and H. Ohtsubo, *Proc. Natl. Acad. Sci. USA*, 1978, **75**, 5983.

37 L.F. Povirk, W. Kühnlein and F. Hutchinson, *Biochim. Biophys. Acta*, 1978, **521**, 126.

38 C.A.G. Haasnoot, U.K. Pandit, C. Kruk and C.W. Hilbers, *J. Biomol. Struct. Dynam.*, 1984, **2**, 449.

39 M.A. Akkerman, C.A.G. Haasnoot and C.W. Hilbers, *Eur. J. Biochem.*, 1988, **173**, 211.

40 M.A. Chien, A.P. Grollman and S.B. Horwitz, *Biochemistry*, 1977, **16**, 3641.
41 L.F. Povirk, M. Hogan and N. Dattagupta, *Biochemistry*, 1979, **18**, 96.
42 C.-H. Huang, L. Galvan and S.T. Crooke, *Biochemistry*, 1980, **19**, 1761.
43 J. Kuwahara and Y. Sugiura, *Proc. Natl. Acad. Sci. USA*, 1988, **85**, 2459.
44 Y. Iitaka, H. Nakamura, T. Nakatani, Y. Muraoka, A. Fujii, T. Takita and H. Umezawa, *J. Antibiot.*, 1978, **31**, 1070.
45 R.M. Burger, S.B. Horwitz, J. Peisach and J.B. Wittenberg, *J. Biol. Chem.*, 1979, **254**, 12 299.
46 R.M. Burger, J. Peisach and S.B. Horwitz, *J. Biol. Chem.*, 1981, **256**, 11 636.
47 R.M. Burger, T.A. Kent, S.B. Horwitz, E. Münck and J. Peisach, *J. Biol. Chem.*, 1983, **258**, 1559.
48 J. Sam, X. Tang and J. Peisach, *J. Am. Chem. Soc.*, 1994, **116**, 5250.
49 T.E. Westre, K.E. Loeb, J.M. Zaleski, B. Hedman, K.O. Hodgson and E.I. Solomon, *J. Am. Chem. Soc.*, 1995, **117**, 1309.
50 R.M. Burger, J. Peisach and S.B. Horwitz, *J. Biol. Chem.*, 1982, **257**, 3372.
51 J. Templin, L. Beery, S. Lyman, R.W. Byrnes, W.E. Antholine and D.H. Petering, *Biochem. Pharmacol.*, 1992, **43**, 615.
52 D.H. Petering, W.E. Antholine and L.A. Saryan, in *Anticancer and Interferon Agents. Drugs and the Pharmaceutical Sciences*, ed. R.M. Ottenbrite and G.B. Butler, Marcel Dekker, New York, 1984, Volume 24, p. 203.
53 W.E. Antholine and D.H. Petering, *Biochem. Biophys. Res. Commun.*, 1979, **91**, 528.
54 W.E. Antholine, D.H. Petering, L.A. Saryan and C.E. Brown, *Proc. Natl. Acad. Sci. USA*, 1981, **78**, 7517.
55 J.P. Albertini and A. Garnier-Suillerot, *Biochemistry*, 1984, **23**, 47.
56 P. Fulmer, C. Zhao, W. Li, E. DeRose, W.E. Antholine and D.H. Petering, *Biochemistry*, 1997, **36**, 4367.
57 R.X. Xu, W.E. Antholine and D.H. Petering, *J. Biol. Chem.*, 1992, **267**, 944.
58 C.-H. Chang and C.F. Meares, *Biochemistry*, 1984, **23**, 2268.
59 R.X. Xu, W.E. Antholine and D.H. Petering, *J. Biol. Chem.*, 1992, **267**, 950.
60 P. Fulmer and D.H. Petering, *Biochemistry*, 1994, **33**, 5319.
61 M. Chikira, W.E. Antholine and D.H. Petering, *J. Biol. Chem.*, 1989, **264**, 21 478.
62 M. Chikira, K. Sakamoto, W.E. Antholine and D.H. Petering, unpublished information.
63 R.X. Xu, D. Nettesheim, J.D. Otvos and D.H. Petering, *Biochemistry*, 1994, **33**, 907.
64 W. Wu, D.E. Vanderwall, S.M. Lui, X.-J. Tang, C.J. Turner, J.W. Kozarich and J. Stubbe, *J. Am. Chem. Soc.*, 1996, **118**, 1268.
65 Q. Mao, P. Fulmer, W. Li, E.F. DeRose and D.H. Petering, *J. Biol. Chem.*, 1996, **271**, 6185.
66 W. Wu, D.E. Vanderwall, C.J. Turner, J.W. Kozarich and J. Stubbe, *J. Am. Chem. Soc.*, 1996, **118**, 1281.
67 D.E. Vanderwall, S.M. Lui, W.Wu, C.J. Turner, J.W. Kozarich and J. Stubbe, *Chem. Biol.*, 1997, **4**, 373.
68 B.J. Carter, V.S. Murty, K.S. Reddy, S.-N. Wang and S.M. Hecht, *J. Biol. Chem.*, 1990, **265**, 4193.
69 J. Kross, W.D. Henner, S.M. Hecht and W.A. Haseltine, *Biochemistry*, 1982, **21**, 4310.
70 L. F. Povirk, M. Hogan, M. Buechner and N. Dattagupta, *Biochemistry*, 1981, **20**, 665.
71 N. Hamamichi, A. Natrajan and S.M. Hecht, *J. Am. Chem. Soc.*, 1992, **114**, 6278.
72 S.A. Kane, A. Natrajan and S.M. Hecht, *J. Biol. Chem.*, 1994, **269**, 10 899.
73 F. Arcamone, *Cancer Res.*, 1985, **45**, 5995.

74 J.W. Lown, *Mol. Cell. Biochem.*, 1983, **55**, 17.
75 G. Gaudiano and T.H. Koch, *Chem. Res. Toxicol.*, 1991, **4**, 2.
76 B. Sinha, M.A. Trush, K.A. Kennedy and E.G. Mimnaugh, *Cancer Res.*, 1984, **44**, 2892.
77 H. Bernaldo, A. Garnier-Suillerot, L. Tosi and F. Lavelle, *Biochemistry*, 1985, **24**, 284.

78 S.A. Murphree, T.R. Tritton, P.L. Smith and A.C. Sartorelli, *Biochim. Biophys. Acta*, 1981, **649**, 317.
79 E.J.F. Demant, *Eur. J. Biochem.*, 1984, **142**, 571.
80 H.-J. Wang, G. Ughetto. G.J. Quigley and A. Rich, *Biochemistry*, 1987, **26**, 1152.
81 D.P. Remeta, C.P. Mudd, R.L. Berger and K.J. Breslauer, *Biochemistry*, 1993, **32**, 5064.
82 S.-S. Pan, L. Pedersen and N.R. Bachur, *Mol. Pharmacol.*, 1981, **19**, 184.
83 E. Kukielka and A.I. Cederbaum, *Arch. Biochem. Biophys.*, 1990, **283**, 326.
84 H. Eliot, L. Gianni and C. Myers, *Biochemistry*, 1984, **23**, 928.
85 L. Gianni, L. Vigano, C. Lanzi, M. Niggeler and V. Malatesta, *J. Natl. Cancer Inst.*, 1988, **80**, 1104.
86 G. Minotti, *Arch. Biochem. Biophys.*, 1990, **277**, 268.
87 K. Mailer and D.H. Petering, *Biochem. Pharmacol.*, 1976, **25**, 2085.
88 D. Gelvan, E. Berg, P. Saltman and A. Samuni, *Biochem. Pharmacol.*, 1990, **39**, 1289.
89 K.M. Tewey, R.C. Rowe, L. Yang, B.D. Halligan and L.F. Liu, *Science*, 1984, **266**, 466.
90 A.H. Corbett and N. Osheroff, *Chem. Res. Toxicol.*, 1993, **6**, 585.
91 G. Powis, in *The Toxicity of Anticancer Drugs*, ed. G. Powis and M.P. Hacker, Pergamon Press, New York, 1991, p. 106.
92 J.H. Doroshow, C. Tallent and J.E. Schechter, *Am. J. Pathol.*, 1985, **118**, 288.
93 R.D. Olson and P.S. Mushlin, *FASEB J.*, 1990, **4**, 3076.
94 H. Porumb and I. Petrescu, *Prog. Biophys. Molec. Biol.*, 1986, **48**, 103.
95 P. Alderton, J. Gross and M.D. Green, *Cancer Res.*, 1990, **50**, 5136.
96 J.L. Speyer, M.D. Green, E. Kramer, M. Rey, J. Sanger, C. Ward, N. Dubin, V. Ferrans, P. Stacy, A. Zeleniuch-Jacquotte, J. Wernz, F. Feit, W. Slater, R. Blum and F.M. Muggia, *N. Engl. J. Med.*, 1988, **319**, 745.
97 B.B. Hasinoff, *Agents Actions*, 1989, **26**, 378.
98 B.B. Hasinoff and S.V. Kala, *Agents Actions*, 1993, **30**, 3972.
99 S. Nyayapati, J. Xiao, R.W. Byrnes, F. Lornitzo, A. Quesada and D.H. Petering, Abstracts of the Meeting of the American Society for Biochemistry and Molecular Biology, *FASEB J.*, 1994, **8**, A1467, Abstract 1212.
100 S. Nyayapati, G. Afshan, F. Lornitzo, R.W. Byrnes and D.H. Petering, *Fr. Rad. Biol. Med.*, 1996, **20**, 319.
101 L. Gianni, L. Vigano, C. Lanzi, M. Niggeler and V. Malatesta, *J. Natl. Cancer Inst.*, 1988, **80**, 1104.
102 G. Minotti, *Arch. Biochem. Biophys.*, 1990, **277**, 268.
103 D. Gelvan and A. Samuni, *Cancer. Res.*, 1988, **48**, 5645.
104 S.S. Massoud and R.B. Jordan, *Inorg. Chem.*, 1991, **30**, 4551.
105 G.F. Vile, C.C. Winterbourn and H.C. Sutton, *Arch. Biochem. Biophys.*, 1987, **259**, 616.
106 C.E. Thomas and S.D. Aust, *Arch. Biochem. Biophys.*, 1986, **248**, 684.
107 G. Minotti, *Arch. Biochem. Biophys.*, 1989, **268**, 398.
108 E.J.F. Demant, *FEBS Lett.*, 1984, **176**, 97.

109 E. Kukielka and A.I. Cederbaum, *Arch. Biochem. Biophys.*, 1990, **283**, 326.
110 L. Gianni, J.L. Zweier, A. Levy and C.E. Myers, *J. Biol. Chem.*, 1985, **260**, 6820.
111 B.B. Hasinoff, *Biochem. Cell Biol.*, 1990, **68**, 1331.
112 J. Muindi, B.K. Sinha, L. Gianni and C. Myers, *Mol. Pharmacol.*, 1985, **27**, 356.
113 H. Eliot, L. Gianni and C. Myers, *Biochemistry*, 1984, **23**, 928.
114 T. Miura, S. Muraoka and T. Ogiso, *Pharmacol. Toxicol.*, 1994, **74**, 89.
115 C. Myers, L. Gianni, C.B. Simone, R. Klecker and R. Greene, *Biochemistry*, 1982, **21**, 1707.
116 J. Xiao and D.H. Petering, to be submitted for publication.
117 D.H. Petering and H.G. Petering, in *Antineoplastic and Immunosuppressive Agents, Handbook of Experimental Pharmacology 38*, ed. A.C. Sartorelli and D.G. Johns, Springer-Verlag, Berlin, 1975, Vol. 2, p. 841.
118 K.C. Agrawal and A.C. Sartorelli, in *Antineoplastic and Immunosuppressive Agents, Handbook of Experimental Pharmacology 38*, ed. A.C. Sartorelli and D.G. Johns, Springer-Verlag, Berlin, 1975, Vol. 2, p. 793.
119 D.H. Petering, in *Metal Ions in Biological Systems*, ed. H. Sigel, Marcel Dekker, Basel, 1980, Vol. 2, p. 197.
120 D.A. Winkelmann, Y. Bermke and D.H. Petering, *Bioinorg. Chem.*, 1974, **3**, 261.
121 D.T. Minkel, L.A. Saryan and D.H. Petering, *Cancer Res.*, 1978, **38**, 124.
122 L.A. Saryan, E. Ankel, C. Krishnamurti and D.H. Petering, *J. Med. Chem.*, 1979, **22**, 1218.
123 I.H. Krakoff, E. Etcubanas, C. Tan, K. Mayer, V. Bethune and J.H. Burchenal, *Cancer Chemother. Rep. (Pt 1)*, 1974, **58**, 207.
124 W.E. Antholine, J.M. Knight and D.H. Petering, *Inorg. Chem.*, 1977, **16**, 569.
125 L.A. Saryan, K. Mailer, C. Krishamurti, W. Antholine and D.H. Petering, *Biochem. Pharmacol.*, 1981, **30**, 1595.
126 R.W. Byrnes, M. Mohan, W.E. Antholine, R.X. Xu and D.H. Petering, *Biochemistry*, 1990, **29**, 7046.
127 D.S. Sigman and C-h.B. Hong, *Metal-DNA Chemistry*, ed. T. Tullius, American Chemical Society, Washington, DC, 1989, p. 24.
128 R.W. Byrnes, W.E. Antholine and D.H. Petering, *Fr. Rad. Biol. Med.*, 1992, **12**, 457.
129 R.W. Byrnes, W.E. Antholine and D.H. Petering, *Fr. Rad. Biol. Med.*, 1992, **13**, 469.
130 L.G. Sillén and A.E. Martell, *Stability Constants, Supplement No. 1*, The Chemical Society, London, 1971, p. 717.
131 M. Gokey, S. Logan, T. Blumenfield, Q. Mao and D.H. Petering, to be submitted for publication.

Subject Index